**Texts and Monographs in
Economics and Mathematical Systems**

Edited by
Martin J. Beckmann and Wilhelm Krelle

Claude Ponsard

History of Spatial Economic Theory

With 44 Figures and 752 References

Springer-Verlag
Berlin Heidelberg New York Tokyo 1983

Claude Ponsard
Professor at the University of Dijon and
Director of the Institute of Economic Mathematics (C.N.R.S.)
4, boulevard Gabriel
F-21000 Dijon

Translated into English by
Benjamin H. Stevens
President of the Regional Science Research Institute

Margaret Chevaillier
Lecturer at the University of Dijon

Joaquin P. Pujol
Associated with the International Monetary Fund

Revised version of the original French edition entitled
"Histoire des théories économiques spatiales"
Paris, Librairie Armand Colin, 1958

ISBN 3-540-12802-6 Springer-Verlag Berlin Heidelberg NewYork Tokyo
ISBN 0-387-12802-6 Springer-Verlag NewYork Heidelberg Berlin Tokyo

This work is subject to copyright. All rights are reserved, whether the whole or part of materials is concerned, specifically those of translation, reprinting, re-use of illustrations, broadcasting, reproduction by photocopying machine or similar means, and storage in data banks. Under § 54 of the German Copyright Law where copies are made for other than private use, a fee is payable to "Verwertungsgesellschaft Wort", Munich.

© Springer-Verlag Berlin Heidelberg 1983
Printed in Germany

The use of registered names, trademarks, etc. in this publication does not imply, even in the absence of a specific statement, that such names are exempt from the relevant protective laws and regulations and therefore free for general use.

Printing: J. Beltz, Hemsbach. Binding: J. Schäffer OHG, Grünstadt
2142/3140-543210

This volume is dedicated to scholars all over the world who are now contributing to the future history of spatial economic theory.

For our science, finally, the question how the economy fits into space not only opens a new field but leads in the final analysis to a new formulation of the entire theory of economics.

August Lösch

Acknowledgements

It has taken a long time to put this book together and I am greatly indebted to a number of people.

First of all, Mr. Walter ISARD, Professor at Cornell University, heartily encouraged the idea of an English edition of my "Histoire des théories économiques spatiales". Mr. Benjamin H. STEVENS, President of the Regional Science Research Institute, and his associate, at that time, Mr. Joaquin P. PUJOL produced the first translation of the 1958 French edition. My gratitude goes out to my American colleagues whose contribution was decisive.

More recently that version has been revised, enlarged and brought up to date in France. I wish to thank Mrs. Margaret CHEVAILLIER, Lecturer at the University of Dijon, who took on the difficult task of translating my additions to the book and of co-ordinating the work as a whole. Miss Marie-Christine NESME, Secretary of l'Institut de Mathématiques Economiques, undertook the particularly thankless work of typing the manuscript. I am deeply grateful for her unfailing dedication.

Of course, I alone remain responsible for this translation which I have followed through its different stages checking all the typescripts.

Finally, I wish to particularly thank, Mr. Martin J. BECKMANN, Professor at Brown University and Technische Universität München, who agreed to publish this book in the new founded collection "Texts and Monographs in Economics and Mathematical Systems, Springer-Verlag".

In the preface to the original French edition, I paid tribute to those masters who had given me unspairing encouragement. I remain as indebted to them now as I did then and take this opportunity to renew my gratitude.

Claude PONSARD

Contents

Introduction.	1
Chapter 1. Before Thunen.	5
Section 1. Space in 18th Century Economic Thought.	6
Section 2. The Break Between Spatial Analysis and the English Classical School.	10
Chapter 2. Johann-Heinrich Von Thunen.	13
Introduction. Genesis of Thunen's Theories.	13
Section 1. The Model of Concentric Rings.	14
Section 2. Thunen and His Time.	16
Conclusion.	17
Chapter 3. From Thunen to Weber.	19
Section 1. The Anglo-Saxon Descriptions.	19
Section 2. Towards the First German Deductive Theories.	20
Chapter 4. Alfred Weber.	23
Introduction. Historical and Theoretical Background of the Weberian Analysis.	23
Section 1. Parameters and Assumptions.	24
Section 2. The Three Types of Location Orientation.	25
1. The Point of Minimum Transport Cost.	25
2. Labor Orientation.	26
3. Agglomeration.	27
4. Total Orientation.	29
Section 3. The System of Locations.	30
Conclusion.	31
Chapter 5. From Weber to Palander.	33
Section 1. Andreas Predohl.	33
1. Substitution and Location.	34
2. The General Theory of Location.	35
Section 2. From Englander to Christaller.	36
Section 3. From Fetter to Ohlin.	41
Chapter 6. Tord Palander.	47
Introduction. The Goals and Methods of Palander's Theories.	47
Section 1. Market and Location.	50
Section 2. Transport Cost, Location of Production, and Market Size.	53
Section 3. Conditions of Competition and Location.	57
Conclusion.	60
Chapter 7. From Palander to Losch.	61
Chapter 8. August Losch.	65
Introduction. The Background for Losch's Theories.	65
Section 1. The Theory of Location.	66
Section 2. The Theory of Regions.	72
Section 3. The Theory of Exchange.	80
Conclusion.	87
Chapter 9. From Losch to the Nineteen Fifties.	89
Section 1. Partial Spatial Equilibria.	89
Section 2. General Spatial Equilibrium.	92

Chapter 10. Since the Nineteen Fifties. ... 99
    Introduction. The Scientific Environment During the Second Half of the 20th Century. ... 99
    Section 1. The Paradigms. ... 100
        1. The Perennity of Thunen's Concentric Rings. ... 100
        2. Weber's Problem Generalized. ... 102
        3. Hotelling's Law. ... 106
        4. The Christaller-Losch Central Places Archetype. ... 108
    Section 2. New Directions. ... 113
        1. The Construction of Models of Spatial Interaction. ... 113
        2. The Development of the Theory of General Spatial Equilibrium. ... 116
        3. The Elaboration of a Theory of Spatial Public Economics. ... 117
        4. The Birth of Spatial Econometrics. ... 119
        5. Furthering the Concept of Economic Space. ... 120
    Conclusion. ... 125

Conclusion. ... 127

Mathematical Appendix. ... 129
I. Von Thunen's Models. ... 131
    1. Concentric Rings. ... 131
    2. Theory of the Natural Wage. ... 134
II. Weber's Models. ... 136
    1. The Point of Minimum Transport Cost. ... 136
    2. Labor Deviation (The Isodapane Technique). ... 143
    3. Agglomeration. ... 143
III. Palander's Models. ... 148
    1. The Delineation of the Market. ... 148
    2. Price Policy and Transport Costs. ... 151
    3. Isoline Technique. ... 158
    4. Law of Refraction. ... 165
    5. Conditions of Competition and Location. ... 167
IV. Losch's Models. ... 175
    1. Location of Two Agricultural Products. ... 175
    2. The Equations of General Spatial Equilibrium. ... 182
    3. Economic Regions. ... 185
    4. Spatial Differentiation of Prices. ... 191

Chronological Bibliography. ... 195
Author Index. ... 231

Introduction

The concept of space has always been a fundamental element in various branches of knowledge. The concept often appears in the evolution of knowledge, either as a basis of theory or as a factor in research. It is associated, more or less directly, with all the history of scientific thought.

At the level of simple common sense, the importance of the concept of space is only equaled by its lack of precision. It was part of legend before becoming part of history. To indicate the founding of Rome, Romulus started by drawing the boundaries, locating its landmarks in a discontinuous space after having cut the limits of a continuous space. However, neither geographical explorations nor mathematico-logical speculations have ever completely removed the mystery from the concept of space. For all its simple common sense, its mystique remains intact.

The privileged position occupied by the concept of space in the history of science and the vagueness of its meaning in the current use of the term, far from constituting a paradox, are mutually explanatory. Every concept of space is necessarily the result of an abstraction, whether the process by which it is reached is through mathematics, psychology, biology, or any other discipline. At the level of common knowledge, the space-time concept is the base upon which are arranged individual experiences. It is thus easy to understand how the concept of space can be understood only through an orderly arrangement of these experiences and their integration into a logical scheme.

With regard to economic science, the concept of space has raised similar difficulties. The backwardness of economic knowledge in the area of space becomes apparent upon realization that by the second half of the 20th Century, its study was just beginning to interest a significant number of economists, after having long remained the preoccupation of a few of the less known economic theorists. If the rapid development of both economics in general and economic dynamics in particular is taken into account, the disparity in the results achieved in spatial economics can be seen as the result of the unequal emphasis given to various areas of economic knowledge.

Two major characteristics of the history of spatial economic theories have been less development of the concept of economic space than the concepts of space used by other sciences, and less development of economic analysis of spatial phenomena than other economic phenomena, especially dynamics. The difficulty that economics had in defining its own objectives as a science is related to its difficulty in liberating itself from other forms of thought, especially of a moral or political nature. This is why initial attempts at spatial economics are not found until the beginning of the 19th Century. A controversy could, no doubt, be started as to whether or not some beginnings of spatial economic thought are found in the literature of the 18th Century. Whatever the cause, the development of a true economic consideration of spatial phenomena occurred in the 19th and particularly in the 20th Century. Its history, therefore, is relatively recent.

On the other hand, the little attention paid by economists to the spatial factor explains the continuous divergence between progress in spatial economics and progress in general economics. Thus, the history of spatial theories is not only recent, but also brief. The difficulties that may exist in mastering it are due more to the novelty of the subject than to the true scope of the field covered by it. It is not easy to gather and interrelate the contributions in the field since they are dispersed all over the world. Even more difficult is the task of penetrating the thoughts of the authors who were led, either by necessity or by imprecision of thought, to reason with the help of vague and inappropriate concepts, or to express themselves in a confused language or with the help of a complicated mathematical apparatus.

The history of spatial theories is that of the progressive refinement of the frameworks of analysis, the patient search for their coordination and their unity, and the slow development of appropriate descriptive and analytical methods. It is to be understood that these advances are not always continuous and wellordered, even if the general direction of development is well marked and relatively simple to underline (1).

The principle task and main difficulty of the historian of spatial economic analysis is not merely to describe the content of writings, but to discover the general lines of the authors'thoughts, to be able to isolate the important schools of thought, and to trace the connections which unite the major authors and tie the less important authors to them. Through time, the degree of economic knowledge of the spatial factors varies from one author to another, from one school of thought to another, from one country to another, so that it is often arduous to disect the currents of research that have been exhausted from those that have not yet delivered all their possible results. At each step it is also important to disclose not only the immediate contributors but also the forerunners of the future, since the intuition of genius is often distinct from systematic elaboration. From discontinuous progress, diverse contributions, trial and error, it is especially important to distill the continuity of a homogeneous thought through its object and the constancy of continuing study by its objective. With respect to the latter point, the difference in points of view should not overshadow the essential unity of the development of spatial thought.

Certainly, these difficulties are not unique to the understanding of the history of spatial economics. They are present in any history of theory, whether economic or not. Also, as with any other history of thought, the history of spatial economic theories appears a priori to be dominated by a few major authors. Their works appear as a vast creative synthesis - a synthesis in that it reflects the knowledge acquired at a given time - a creative synthesis in that by its organization and content it marks a radical expansion of knowledge with respect to what was previously known. These works thus provide the historian with the first indispensable reference points. Starting with them, it becomes possible to find a basis for order and a non-arbitrary organizing principle. Around them, it becomes possible to place the works of secondary authors, who are important because of their marginal contributions or because of the intellectual atmosphere they create.

Among the several important authors who marks the critical moments of the evolution of spatial economic theories, Thunen, although he had already made a first creative synthesis, is still thought of as a pioneer. The essentials of his framework can be easily found in previous works, notably in some manuals on agricultural economics of the 18th Century. His strong originality, however, rests in his treatment of the topic. Next, Weber's theory of industrial location displays the critical influence of pure economics in the spatial domain. Then, with Predohl, a decisive beginning is made toward a general equilibrium theory of location, which was then followed by Palander and Losch. The combined efforts of these authors cover almost one hundred and fifty years of the history of spatial theory and give continuity to a collection of extremely diverse contributions. The minor currents of thought which linked these authors together are rarely simple. Even though their works reflect the influence of the main works, they did make certain minor contributions which had a considerable influence in their own time and could regain importance today.

Having found the path through which the spatial economic theories developed, there is still to be found the place of these theories among the whole economic field. This is as important to the understanding of the inception of spatial theories as are the links that assured its development. The fact that spatial analysis developed outside traditional economic theory explains the unorthodox character of spatial theories until the middle of the 20th Century. The result is that, at first glance, their history seems perfectly autonomous, without relation to the history of orthodox economics. This impression is reinforced by the fact that the theories were the work of specialists. The attempts of traditional economists, such as Ohlin, to include the spatial factor within that discipline appear as exceptions to this general rule and resulted in failure.

However, the absence of influence is only apparent. On the contrary, very precise relationships never ceased to exist between spatial theories and other economic theories. General economics constantly exerted an effect of intellectual domination on spatial economic analysis, since its influence was always asymetrical and irreversible. Thunen cannot be understood without comparing him with Adam Smith and Ricardo. Weber's mechanistic theory is a projection of pure economics on the spatial domain. Predohl's theory and all the later development of a general equilibrium theory of location show the influence of the Lausanne School. All the neo-classical price theories served as an impetus to the works of Palander and Losch. Similarly, authors such as Hicks and Leontief influenced the work of Isard. The examples are numerous.

However, it would be a great error to conclude that spatial economics has only introduced a supplementary dimension into the contemporary general economic models. The effect of the influence exerted by these models is only one element determining the history of spatial economics; the evolution of the events and the accumulation of spatial knowledge itself are two other important factors. It would be impossible to account for the autonomy of spatial research if these factors were ignored. To underestimate them would result in a false concept of the historical and logical relationship between spatial theories and general economic theories. Historically, the evolution of the relationship constitutes a fundamental aspect of the development of spatial theories. Such an evolution would be incomprehensible if spatial analysis had only been the enlargement of general economics to incorporate a new dimension. On the contrary, it is to the extent that spatial economic analysis kept a certain autonomy and progressed as a function of accumulated results that it was able to control its connections with general economics.

The history of these connections is not simple. In the beginning, the first spatial theories showed more points of departure than of coordination. Later, the tendency arose to think that location theories could be deduced from general economic theory through specification of appropriate assumptions. Finally, the idea became prevalent that, on the contrary, spatial economics embraces punctiform economics as a special case where the spatial configurations are explicitly or implicitly ignored. That is, while historically spatial economic analysis was dominated by traditional economic analysis, at the end of its evolution it has become logically more general. Its history, therefore, will be entwined with the history of this discovery and its proof.

---

(1)For a very brief synthesis, cf.Trias-Fargas (R.) [267],Flores (E.) [269].

Chapter 1. Before Thunen

Any historian who attempted to look back into Ancient Times to find the forerunners of spatial economic analysis would be setting out on a dangerous venture and would be laying himself open to two risks. Firstly that of betraying certain doctrines and arbitrarily uncovering a significance which had remained oblivious to the authors. Secondly, of undermining the concept of economic space to the point of depriving it of its abstract nature. This in turn could lead to people wrongly attributing to spatial economics certain well-founded concrete notions of economic or political geography.

The learned works by Dockes [404] and Dandri [672] confirm the idea that we have to wait until the 18th Century to find the first insight into the abstract concept of economic space (1).

It is true that as early as the 16th Century, economic literature discussed a number of constant preoccupations such as rural/urban relationships, the location and size of urban centers, industries and farming areas, the role and importance of means and routes of communication, domestic and foreign trade patterns, the geographical distribution of populations and wealth etc. Dockes showed how a double current kept the history of economic thought moving from the 16th to 18th Century. On the one hand, both in France and England, mercantilist doctrines advocated the concentration of production which lead to an unequal geographical distribution of mean and wealth: this distribution being not only accepted but also encouraged. Moreover, we know this to be also true of German mercantilism. 18th Century liberal doctrines were, on the contrary, in favour of greater equilibrium and relative equality in the distribution of economic activities as a result of their dispersal. With all the reserve necessary when interpreting a still unmethodical thought, one might say that mercantilism and liberalism differ fundamentally on two counts. Mercantilism favours a voluntary geographic distribution where concentration is judged to be in accordance with the good of the State, while Liberalism advocates a spontaneous geographic distribution where dispersal is the fruit of the natural order as well as the rational order of society. On the other hand, however, in addition to these two different standpoints, a current of thought emerged which lead to a more abstract and rigorous understanding of spatial relationships. As time goes on there is a movement away from fragmentary views offering only incomplete observations, towards numerous advances which, based on sound knowledge and reasoning, were to explicitly introduce the spatial factor. This movement culminated in several works of synthesis in which the role and importance of the spatial factor were, at last, fully recognized.

It follows that the careful reflection sparked off by these opposing standpoints helped scientific analysis to make considerable headway. We are entitled to claim that certain 18th Century authors, both mercantilists and liberals, were indeed the true forerunners of spatial economic analysis in the strict sense of the term.

Section 1. SPACE IN 18th CENTURY ECONOMIC THOUGHT

Cantillon's work beats the mark of the first, and no doubt the most important, forerunner of spatial economics (2).

In his significant work, Cantillon attempts to describe economic patterns [1]. All his thoughts are dominated by the necessity for the economy to regulate its business circuits and trading patterns, which were the common preoccupation of 18th Century authors. His "Essay" is, to begin with, the work of a banker who defines his own ideas in a period preoccupied by monetary debate. Essentially, the work contains an answer to and refutation of the thesis of John Law. It is in this perspective that we can understand the structure of the "Essay" whose composition seems a priori to have no logical organization. With respect to spatial economic analysis, Cantillon shows a superiority over his contemporaries in having recognized not only the vertical circuits of the economy (large payments - small payments), but also the horizontal circuits (urban-rural interdependence). He proceeds to a systematic analysis of the latter. Led by his monetary preoccupations to search for a general concept of the nature of wealth and the process of its formation, Cantillon develops a general explanation of the functioning of the system, integrating the spatial factor. Transport economics explain the birth of villages, towns, cities and a capital, and also the development of credit, as a means of economizing on the transport of money. The whole "Essay" is dominated by a horror of wasteful commercial circuits. This horror was shared by other contemporary economists, but Cantillon was the only one who developed through the spatial factor an explicit analysis of them.

The "Essay" starts with a description of general equilibrium which foreshadows Losch's theory of economic regions. For Cantillon, the whole social organization is subordinated to the fertility of the land, which derives its richness from man's work. This is how the value of wealth and its inequality of distribution are explained. However, economies of time and transportation force the agents of production to locate near the land that they work and exploit. This causes the economically necessary birth of villages. Village dimensions are a function of the number of inhabitants who are employed in the fields and the size of the village hinterland is a function of the nature of the agriculture. Transportation economies transform some villages into towns: that is, market places. The radius of the area of these markets defines the extent of the influence of each town, whose size is a function of the density of the population located in its zone of influence. Similarly, cities originate from the agglomeration of large landowners. The size of the city is a function of its expenditure: that is, size depends on the product of the land belonging to its inhabitants after deducting taxes and expenses of transportation, which reduce the receipts of the land which is located far from the center. Capitals, like cities, are born from the concentration of the larger landowners around the head of state. Contrary to the cities whose locations remain indeterminate, the location of the capital is precise and its proximity to the sea or important rivers plays a strategic role, due to the availability of water transportation.

Thanks to Cantillon's analysis of economic surfaces, the vertical and horizontal circuits of the system become closely associated and their interdependence permits the complete integration of the spatial factor into his other theoretical developments, even though the latter are heavily oriented towards a monetary and banking system. After posing a series of equations relating to comparative labor value of different agents of production, relative value of goods, labor value and land value, Cantillon describes the role of entrepreneurs in the distribution of products and that of landowners whose demand ultimately determines the uses of the land and the variation in prices. However, the interdependence between the vertical circuits (expenditures) and the horizontal ones (locations) is such that the whole circulation of the flows implies a spatial multiplier. The quantitative variations and the qualitative modifications of the demand of landowners determine, together with the multiplier effect, the nature of cultivation, the spatial extension and, finally, the number and location of population.

The whole monetary problem, as posed by Cantillon, is dominated by this functional interdependence between the vertical and horizontal circuits. The originality of Cantillon's work, as compared to other contemporary writings, lies largely in the place that he gives to this interrelation. The explanation of the price level leads him to a shaded "quantitativism", since at each place the nominal price is determined in reference to a market from which the quantities sold on other local markets must be subtracted; and the cost of transporting goods to other local markets must be integrated into the resulting prices. From this we obtain a gradation of prices as a function of the distance from the place of production and, at the same time, the influence that distant markets will have on the prices in the markets under consideration. On the other hand, the quantity of money in circulation depends on the relationship between urban and rural areas, since the different rents, among which the net product of the soil is divided, are distributed among social classes located at different places. But rural and urban behavior differs with regard to consumption expenditures, the velocity of circulation of money, the use of barter transactions, rural home consumption, etc. The entrepreneur is then defined by his function as money and goods distributor in the urban centers. The description of the horizontal circuits of money and real flows led Cantillon to the idea of a regional balance.
On the debit side, he puts the debts of the rural areas to the cities and on the credit side, the returns on the sale of that fraction of the produce which is transported to the cities. He develops similarly a balance between the capital city and the rural and other urban areas and by means of this set of accounts he is able to consider a whole set of flows, mainly fiscal.

The transportation of merchandise and the financial risk that it implies, due to the need for the equilibrium of regional balances, gives birth to local differences in prices. Since prices increase, ceteris paribus, with the distance from the point of production, the only products which will participate in exchange will be those whose sales radius is long enough to reach the urban markets. In this manner, Cantillon foreshadows Thunen's theory of concentric rings and is able to show how the prices at the urban markets determine the distribution of cultivation around the cities, taking into account the modes of transportation and their respective costs. This transportation economy leads Cantillon to point out the advantages of industrial decentralization. In this he reflects a common preoccupation of the Physiocrats. In any case, Cantillon not only foreshadows the Physiocrats but he also extends Mercantilism. His preoccupation with locating manufacturing in the provinces leads him less to elaborate on a theory of industrial location - which he hardly discusses - than to extol the advantages which the provinces will derive in exchange. By shipping manufactured goods, they are able to obtain a greater return on their trade and to correct to their advantage the inequality of the circulation of money in the State. Whatever we make of his Mercantilist views, Cantillon postulates, in an implicit but remarkable way, the identity of interregional and international trade since, in both cases, manufactured products bring in gold under the best conditions and thus result both in the prosperity of the regions and in the power of State through the accumulation of their respective gold stocks. In other words, Cantillon leaves aside the theory of comparative costs. Every exchange mechanism rests on local price differences: that is, ultimately on transport cost, interregional inequality of revenue, and comparative length of the sales radius of different goods. Finally, the urban-rural exchanges appear to be similar to the trade between different countries. This view dominates the whole of Cantillon's work on the variations of the quantity of money in circulation in a country and its effect on consumption and prices. This is why the Mercantilist State must rest its foreign trade on a powerful merchant marine.

Even in his work on pure monetary and banking techniques, Cantillon starts from a spatial framework, which was the foundation of his analysis. From this unity between interregional and international trade, he concludes that the change phenomenon is identical in the two types of transactions and its level depends on the direction and size of the regional and national balances between debits and credits. Similarly, the banks, including the Central Bank, are not independent of the size of the market in which they participate. Beyond a certain critical size of this market, the Central Bank loses its reason for being, according to Cantillon, since it is no longer able to perform a proper role due to its remoteness from the distant participants in the market.

The 18th Century Liberal economists after Cantillon become more active in their defence of the doctrines of equal spatial distribution of wealth. However this element becomes less centrally important and is now only examined in a piecemeal way. It is not until Condillac that the spatial distribution of wealth is again examined in a more systematic manner.

In France, the Physiocrats were laying stress on the origin and circulation of wealth. Yet Quesnay's "Tableau Economique" for example, attaches no importance whatsoever to the location of the groups of economic agents who make up the poles of the circuit between whom incomes flow. It is only insofar as economists moved away from an abstract model towards concrete problems they were forced to comment if only partially, on geographical facts such as the size of towns, rural-urban relationships and means of transport.

Condillac, in his theory of value breaks away from the Physiocrats and he represents a landmark between the contributions of Cantillon and Thunen [3]. Indeed, Condillac starting from a group of assumptions uses a logical deductive method. His governing idea is that the scarcity of goods varies with place, and so space becomes central to his theory of value and trade.

It is through trade — the process by which an individual relinquishes a less-desired surplus than that surplus obtained from his partner - that goods are attracted to those places where their highest value prevails. His description of commercial circuits leads Condillac to study the origin and location of markets. His ideas hark back to those of Cantillon both as regards the development of urban areas, industry and their market areas. The result being that he makes a clear distinction between local and national markets. Finally Condillac contrasts urban-rural circuits with interregional and international circuits. In the former, products and factors of production always move in the same direction whereas in the latter there is a degree of relocation and hence changes in production and commercial patterns.

However just when Liberalism was intensifying its efforts to remove institutional boundaries and barriers to the free movement of men and wealth so mercantilism came up with new arguments. The fruit of the more rigorous analytical method is the great synthesis by Steuart who, in turn becomes one of the true forerunners of spatial economics [2].

Steuart wrote at the beginning of the industrial revolution, that is during the second half of the eighteenth century. His work incorporates spatial effects, i.e.their impact on the location of population and manufactures, the development of urban areas and transportation routes and finally the division of labor between urban and rural areas.

In a primitive economy where people gather the fruit of the land and characterized by the absence of labor and landed property, the population settles where agricultural produce is to be found. However, the existence of landed property implies that only those men actually needed for harvesting remain attached to the land. The residual population can live elsewhere, namely in towns which develop as a function of the volume of agrarian surplus produced by the farming population. The residual population is on the one hand, made up of landowners and rentiers and on the other hand,artisans. The former can choose where they wish to live such that the location of towns is directly dependent on them. Similarly, the State fixes the capital city by concentrating all the administrative services in one place. Artisans, however, who trade what they produce and consume landowners' surpluses must, necessarily, live close to their buyers. In this way, the size of towns is, according to Steuart, a function of the number of landowners, while for Cantillon, it depended on their net income.

Steuart distinguishes between independent artisans and those who join together to form manufactures. This distinction leads him to examine, in a most original way for his time, the location of manufactures. Indeed, the link is broken between a landowner's place of residence and that of the particular fraction of the working population who now depend on merchants rather than consumers. It is true that the proximity of an urban market is a factor in location especially for manufactures which satisfy the needs of a single town. However, since their market is a national one, the manufactures are attracted by sources of energy and raw materials, by places where subsistence prices and wages are low and above all by transport routes which make the access to those different points of attraction and markets easier.

The locations of those manufactures in town, explain the emergence and growth of new industrial urban areas which are quite distinct from the old cities which serve as places of residence for landowners, rentiers and independent artisans.

As towns and capital cities grow they are beneficial to rural areas once the initial difficulties stemming from the migration towards urban agglomeration have been overcome. The incomes circuit between urban and rural areas finds its origin in agricultural surpluses which are marketed. However, the volume of those surpluses depends on the magnitude of the urban monetary demand. In the long run, the larger the town, the wider its supply area.

Finally, the growth in agglomerations postulates the development of communication routes which make urban-rural relationships closer and encourages the division of labour between them.

Steuart who foreshadows Weber in his analysis of the location of manufactures, points towards Thunen by his description of crop areas. Even before the eighteenth century, land under cultivation around a town, was known to be distributed in concentric rings. Steuart, however, gave an accurate presentation of the framework thanks to the empirical observations he had accumulated. In general a town is surrounded by five rings: (1) vegetable gardens; (2) pasture land used for the production of milk, meat and hay; (3) ploughed fields; (4) pasture land used for fattening animals; (5) forests and fallow land. As he did for the location of manufactures, Steuart once again simply lists the factors determining the location of agricultural activities. It is only with Thunen that the different factors are logically co-ordinated thanks to his concept of situation rent. However, as Losch was to point out much later on, Steuart had realized that the order of the rings can be reversed in particular cases, as for instance, in Paris or Padua where he had lived.

In conclusion, one can certainly regret that authors such as Cantillon, Condillac or Steuart were only forerunners of spatial economics. To the extent that their analyses are broader and more comprehensive than Thunen's, one wonders if later progress within spatial economics would not have been quicker if the base of its development had been broader. The slow evolution of the theory of regions is a case in point, as is also the insufficiency of integration of the monetary factor in all spatial analysis. We can say the same of the difficulties in clarifying the fundamental problems of international economics. In a general way, the whole of the economic system which Cantillon and Steuart propose appears richer and more complete than the simplified model of Thunen and his immediate successors. We cannot deny that it was precisely due to the process of abstraction that Thunen was able to reach a first theoretical formulation, but the number of variables that he retained is so limited that later theorists had great difficulty in extending his theory.

Even more, it is regretable that these authors were unknown to later theorists. Certainly their works did not provide a theory of location, as did Thunen's. But the role played by the spatial factor in their writings is such that its integration into economic analysis is more complete.

With Thunen, and after him, a large number of abstract models remain partially punctiform. From this point of view, the history of spatial economic theories is that of the progressive systematic integration of space into economic analysis, and it is precisely from this point of view that such a history will appear most incomplete.

Section 2. THE BREAK BETWEEN SPATIAL ANALYSIS AND THE ENGLISH CLASSICAL SCHOOL

There is unanimous agreement among authors that Thunen was the initiator of spatial economics [5]. He is currently called the "father of location theory" and his work is recognized as the foundation of spatial economic thought. Without doubt Thunen's claims to such a position are multiple and their authenticity cannot be challenged. As the profound analysis of his work will demonstrate, Thunen departs from the economic thought of his time: that is, the English Classical School. The autonomy, which varied through time, of the German thought from Anglo-Saxon ideas is not sufficient to account for this radical break.

Even the differences between Thunen and Smith with respect to land rent theory are minor when we compare them in the context of their respective contributions [4][5]. The economic significance given by Smith to transportation cost is too relevant and the place that he gives it too important for his framework to be called punctiform as compared to Thünen's spatial scheme. Similarly Smith clearly links the degree of division of labor to the volume of the population and the extent of the market for products. The extent of the market for a commodity depends, in turn on transportation routes and on how transportable the particular commodity is. The different values of goods, value, meaning "natural" value and not price, depends on spatial variations in the elements making up the costs of production. In other words, the wages, profits and rents paid to the factors of production. Similarly domestic trade is explained by the emergence of towns and their relationships with their hinterlands. Finally, in his analysis of international relations and in his plea for free-trade, Smith aptly remarks that a degree of protectionism can result from the respective locations of competitive productions.

But, on the other hand, if we compare Ricardo to Thunen, the latter's break with the classical school is obvious. It appears unreducible when Ricardo, by explicitly reducing the differences in location to differences in fertility of the soil, completely eliminates the spatial factor from his framework. Transportation costs are confused with other costs, thereby losing their unique economic significance. Comparative cost theory was, by the same token, to become the logical foundation of the analysis of international trade, as the analysis could no longer be derived from a location theory.

The importance of these debates is essential to the general evolution of all economic thought which was to follow. The 18th Century, which saw the start of the scientific phase of economic thought, always linked empirical verification to previously accumulated observations. This served to prevent divergent conclusions which could result in deep-rooted incompatibilities. This was so even though this period was already characterized by a trial and error development and the contradictions typical of a period of transition. Adam Smith and Thunen were to be the last representatives of this way of thought. Starting with Ricardo, the great majority of the 19th Century economists engaged in abstract analysis. Pure theory and pure observation were to be reunited only during the 20th Century.

Under these conditions, the break between Thunen and Ricardo is the origin of the dichotomy of economic thought which was partially resolved only after a century and a half of evolution of ideas. At the end of the evolution Beckmann passes a severe judgment upon spatial economics when he asserts that most exhaustive spatial statements rest on the idea that the phenomena under study possess a special character, even a logic of their own, and that their analysis requires the employment of a specific theoretical apparatus [236]. He even rejects most location theories as badly integrated amalgams of weak propositions and extremely simplified models. Although his criticism is somewhat tempered by the recognition of the extreme difficulty of the subject of spatial economics, it underlines the unorthodox and diverse character of the analysis.

In a sense, the whole history of spatial economics is that of a constant search by different authors for a general economic principle. The best works of this group are attempts to coordinate the theories of location with general economic analysis. Beckmann is correct in describing Losch's contribution as the major achievement of these attempts and the first landmark on the path towards a modern analysis of the problem, since the notion of equilibrium serves not only as a concept unifying the previous models, but also as the link to general economic theory.

Looking back, however, we find that it was the theoretical divergence between Thunen and Ricardo which led to the incompatibility of the two schools of thought. If the author of the "Isolated State" and that of the "Principles" had only differed in their method of approach, nothing would have warranted their difference in results. However, Thunen's genius was that he was able to work simultaneously at two levels of abstraction. Through the concrete character of his observations, he prolongs the economic thought of the 18th Century; he remains a contemporary of Smith through the importance which he attaches to facts. On the other hand, by the abstract and mathematical character of his theoretical contribution, he foreshadows the 19th Century and can also be considered a contemporary of Ricardo. However, two works based on opposing assumptions can only result in opposite conclusions. When compared to the spatial model of Thunen, Ricardo's model appears typically punctiform. The dichotomy of economic thought which occurred resulted in the loss of unity. The major course of economic thought ignored the spatial factor; in this sense it is post-ricardian. Those economists who recognized the existence of this factor tended to minimize its importance. Such is the case of Marshall. Others who were conscious of it, such as Chamberlin and Ohlin, were led to try in vain to think of it within a classical framework which was inappropriate.

The spatial factor, although at times included in the elaboration of certain special theories, such as rent theory, monopolistic competition, and international trade theory, was never the object of a clear effort of integration into the general theory, as was the time factor. It was only along the lines set by Thunen's work and at the margin of the major current of economic thought that the spatial factor was accounted for. Thus, while the integration of time into economic analysis became one of the central preoccupations, and was to attract the major efforts, of modern analysis, the integration of the spatial factor was the work of those whom Isard calls the "outsiders" of economic science [160].

---

(1) This idea was first defended in **the** original French edition of this book. Cf.Ponsard (C.): Histoire des theories economiques spatiales. Paris, Librairie A.Colin, 1958.
(2) On this matter, see Popescu (O.)[223]. It seems that this Argentinian economist was the first to see Cantillon as a precursor of space economics. See also Spengler (J.J.) [233] and Hebert (R.F.)[727]

Chapter 2. Johann-Heinrich von Thunen

INTRODUCTION . GENESIS OF THUNEN'S THEORIES

The economic revolution which was beginning to influence development in Northern Germany provided the historical background for Thunen's research on the effects of this revolution on agriculture in Jeverland. The sudden appearance of the market economy and the development of cities were posing new problems at that time.

The comparison of two agricultural estates, whose farming Thunen had watched, led to the generation of his theories. The Gross-Flottbek estate, uninfluenced by an outside market, was farmed in the traditional manner, with a natural system of cultivation consisting of a triennal rotation of crops, not leaving the land fallow for a season. The Klein-Flottbek estate, nearer the Hamburg market, felt the impact of a market economy which tied the method of cultivation to the behavior of prices. This led to a system, developed by Baron von Voght and Staudinger, of intensive cultivation of clover, potatoes, and legumes. The natural economy in which it was necessary to obtain the largest gross product "per man" while preserving the fertility of the soil gave way to an exchange economy, in which it was attempted to maximize the net revenue "per unit of surface".

When Thunen later managed his own estate of Tellow in Mecklenburg, he began to realize the important impact of the neighboring market of Rostock on his production pattern. This combination of circumstances stimulated his interest in the problem of spatial economics (1) and became the impetus for Thunen's work [5].

Thunen's outstanding contribution was to pose clearly, and in spatial terms, some of the problems of an exchange economy such as: the influence of cities and businesses on land rent and the prices of goods produced on the farms, the laws of price formation and the effect of urban growth on cultivated land, communication between city and countryside, the effect of parcelling the land and distance from farm to market on revenue, the decline in land value due to poor location, etc.

He was the first to find solutions to some of these problems, thanks to his meticulous observations recorded in his accounts and to certain purely scientific experiments that he undertook.

Bulow shows the conflict in Thunen's time between "natural location" and "economic location"[172]. The choice of these terms seems to depend on questionable criteria of economic measurement. They have, however, the merit of expressing the revolutionary effect that the growth of towns has had on economic space. With the development of an exchange economy and the domination of prices, rent appears. The idea of net revenue, the difference between the cost of production and the sales price, is tied to the appearance of the market economy. However, as Bulow shows, the theory of rent contains the germ of a theory of space, so that the problem of spatial economics arises, historically, with the market economy and its problems of location. Scientifically, the way is also open to spatial studies of other economic systems at other times in history.

Section 1. THE MODEL OF CONCENTRIC RINGS

The data that Thunen gathered from his experiments led to results which were original from both the technical and economic point of view, especially when compared to the investigation of contemporary agriculturists and the English economists.

A word about his agricultural concepts is necessary at this point. Thunen developed a different theory from that of his teacher, Thaer, but he magnified these differences to an extreme that appears to be unfounded. He accused Thaer of inflexibility in his preference for the English methods of farming, using crop rotation (Fruchtwechselwirtschaft). Thunen believed in greater choice with respect to various methods of farming and favored, therefore, "the law of the relative preference of economic systems" advanced by Rodbertus. Thaer, however, protested the practices of the Mecklenburg farmers who applied methods introduced from Holstein in the 18th Century. Roscher characterizes the "law of Thunen" as a theory of intensity [7]. He summarizes his concept of agricultural statics in the formula:
$$E = T Q H K$$
The richness of the soil, R, is put to work by a force, T (Tatigkeit), to give a certain fertility to the land (Fruchtbarkeit). The fertilizer, H, improves the quality, Q, of the soil and releases its richness. The volume of harvest is thus $E = T R$, where $R = Q H$. Therefore $E = T Q H K$, where K is a factor of technology and constant for a given region, while T and Q are constant for a given soil, so that E depends finally on H. The volume of the harvest therefore depends on the quantity of fertilizer.

However, Thunen's experience, interests, and taste for economic research, while limiting the generality of his abstract thinking, gave him a different picture of the economic pattern from that of the Classical English School.

He understood how Adam Smith attempted to combine the Physiocrats' theory of land rent with a theory of the return to the two other factors of production - capital and labor. But Adam Smith defines land rent as the residual that the farmer receives after paying his labor, the costs of cultivation, and the interest on the capital used. The buildings and other improvements are an integral part of the land and do not provide, therefore, any revenue unless the soil is cultivated. Thunen sees this as an "estate rent": that is, revenue from a rented good. He proposes to define land rent as the surplus that appears after the remuneration of all factors of production, including all those things "that can be separated from the land". Thus he takes away from the concept of land rent all content other than the specific revenue to the land.

As defined above, land rent becomes an indicator of the opportunities involved in various available methods of cultivation and in the choice of various products.

Thunen assumes a continuous, isolated plain uniform in fertility and tranport facilities. In the center of the plain is a city serving only as a punctiform market. On this plain, the optimal location of agricultural activities will be a function of land rent, which is to be maximized. The defining element of the system, rent, by its residual character, is dependent on the distance from the market place because of transportation costs to the city. A single price for each product prevails at the urban market. The cost of transportation is proportional to the weight of each product and the distance from the producing farm and determines relative local prices for each product. These relative prices determine, in turn, a series of land rents (2).

Proof of this is based on experimental results obtained at the estate of Tellow. Each activity was recorded in a diary and the cost of each, in money and grain, was determined annually. This provided a basis for the calculation of costs. Starting from available data and from accounting concepts, the argument is conducted in econometric terms. Thunen's work is remarkable for his time, especially his method, which is far different from that of the English classicists.

Using this procedure, Thunen is able to calculate in terms of grain and money the influence on land rent of the volume of production, (and therefore the influence of the fertility of the soil), and of the distance to the market. Different crops and methods of cultivation develop around the central town. From the economic point of view, the interplay of the variations in land use is organized in a series of concentric rings, since homogeneity of the surface is assumed.

The immediate surroundings of the city, within the first ring, are called the area of "free cultivation" because the rent on the land is high and the crop rotation pattern is not determined a priori. In this area market gardening and dairy farming are found. In Thunen's time, vegetables and milk were difficult and expensive to transport relative to their value. The proximity of the city meant that fertilizer could be easily and cheaply obtained, allowing continual cultivation without loss of fertility. The locational advantage of this zone due to access to fertilizers from the city diminishes with distance and, after a certain point, disappears.

A forestry ring follows because of the superior net revenue of wood over other products in this ring. Wood production, however, runs the risk of being inefficiently located because of the amount of time needed to modify wooded land to adjust to changes in demand. The decrease of land rent which corresponds to the distance from the city makes cereals more profitable on the extreme outer edge of the forested ring.

These cereals are produced in the next three rings, which are distinguished by their mode of cultivation. The first of these three rings uses 2-crop cultivation; the second, pastoral rotation, and the third, 3-crop rotation. Past the limit of these rings, the land rent for grain is negative and consequently the cultivation of cereals stops. The outer boundary of the last ring is then the marginal production location. Thunen is thus led to a "marginalist" theory of price. By this theory the price of the grain must be high enough to pay the highest costs of growing and delivery of the farm, the cultivation of which is still indispensable for the satisfaction of demand, without the land rent becoming negative. Ceteris paribus, an accidental drop in price would diminish the amount of land under cultivation, causing a scarcity and a corresponding rise in price.

In the same way, a modification of the parameters (variations of production or consumption) enlarges or contracts the areas under cultivation until a stable price is determined. This price becomes the equilibrium price which just calls forth the equilibrium production in each ring.

Thunen's marginal price theory has, in its framework, a universal application. The relationship that it establishes between the price and the marginal land rent has a general character. It is also valid for livestock, for example, which occupies the sixth and last ring surrounding the city.

This model of concentric rings suffers from distortions when Thunen reintroduces the factors abstracted earlier for the purpose of simplification. The existence of a waterway that provides a cheaper means of transportation lengthens his rings into a spindle-shaped curve along its banks; the existence of several neighboring cities, each the center of a concentric system of rings, would result in the intersection of some of these rings; the existence of taxes heavy enough to eliminate land rents could distort or destroy some of the rings. Thunen incorporates these effects through a change in the local price. Fertility, which may also distort the ring system, is incorporated not by prices, but by returns and following changes in the width of the rings.

Section 2. THUNEN AND HIS TIME

Some commentators on Thunen's work see only a theory of agricultural location. Certainly his structure remains essentially an agricultural economic geometry, but it has wider application.

It would be fruitless to make an exact account of the variables that are not included in Thunen's theory. Thunen left a simplified model of the German agricultural economy of his time. Although knowledge of his theoretical contribution was not wide spread until later, his work had considerable scientific and political influence within Germany. His disciples were numerous, although they tended to imitate rather than surpass him; and in 1829, borrowing from his arguments, the "law of reunion" was passed by the Prussian government.

This success shows that Thunen was in touch with the spirit of his time, despite his independence and originality [174].

Eighteenth Century German Cameralism had still enough vigor to give Thunen a background of ideas based on natural law, a background that he was later to transcend but not without borrowing from it. The Nineteenth Century, however, was to be the century of positivism and economic and social doctrines incompatible with Cameralism.

This is why Thunen searched for the natural wage, meaning a reasonable wage which would fulfill both a moral and a social obligation (Thunen had read Hegel). The problem of the natural wage, unsolved by Thunen's contemporaries, was examined by the author of the Isolated State by applying a scientific method. This method, it must be emphasized, was revolutionary in its application to such a problem at that time (3).

While maintaining the assumptions of the Isolated State, Thunen asks whether higher wages are possible on the marginal farms. The revenue of these farms covers only the cost of production. A rise in wages would make rent negative. Production would stop, the cultivated surface would contract, and the unemployed labor force would move closer to the center. There, the increased competition with those already employed would cause a drop in wages. The results of this process would mean a loss imposed upon the laborers.

But Thunen maintains that this mechanism may not come into play if the interest rate falls. The assumption of a constant rate of interest must be rejected and the interdependence of wages and interest rates must be considered. To do this, Thunen tries to relate the efficiency of capital to labor by a system of equations (4).

He shows mathematically that the gains from one year's real wages are to the gains of one unit of capital as 1 to z, z representing the real interest rate, measured in rye. Then he concludes that the entrepreneur will be led to increase, through substitution (5), the amount of capital, so that the cost of the work performed by the capital and that of the work performed by the laborers will be in direct proportion to their respective efficiencies of production.

In the Isolated State, the workers have the choice of working for someone else or for themselves on marginal land. If a certain number of workers, agree to put a marginal farm into production and hire other workers, the total mass of workers will be divided into two groups, A and B. Group A will be employed on the original farms and this surplus furnishes the means of subsistence to Group B. Finally Thunen shows that capitalists and workers have a common interest in maximizing a function which relates the surplus of Group A, invested for interest, to the income of the capitalists of Group B.

Moore, writing at a time when the Marshallian concept of substitution had been elaborated [18], understood the lack of precision of Thunen's model. His z factor expresses the relative efficiency of labor and capital only at the "margin of indifference" for substitution between these two factors. It does not express the total efficiency of capital. Therefore Thunen could not in this way draw firm conclusions about the insufficiency of the Ricardian theory of wages at a time when wages were not fair.

If Thunen's theory of wages is placed in opposition to Ricardo's, we can see that Thunen's work overlooks other conclusions that might result from considering the effects of a rise in population. Ricardo shared with Malthus a pessimism that was caused by the unfavorable nature of the relationship between population and cultivated land.

From his correspondence, we know that Thunen had studied the work of Malthus, even though he does not mention it in his book [172]. If population grows beyond the capacity of the land to feed it, there are three possible solutions: limitation through continence, intensification of agricultural cultivation, or intensification of the industrial export economy (as defined by Friedrich List). The latter has no place in Thunen's work, leaving only the first two alternatives. Without doubt, Thunen was malthusian in his acceptance of the existence of a population problem, but he was not pessimistic since he shows that with higher prices for cereals, more intensive and, therefore, more productive cultivation will result.

Thunen, who had also read Lorenz von Stein (6), was inclined to look for a social doctrine in some form of socialism.

CONCLUSION

Despite the limitations and imperfections of his model, Thunen's work had the merit of being the origin of a series of investigations in spatial economics. These investigations were to remain under his influence for a long time.

A pioneer in spatial theory, he was, in addition, the first to reason in marginalist terms, although he did not use such terminology. He was a forerunner in developing the principle of substitution and the problem of imputation.

His works on natural wage are a link — generally forgotten — between Ricardo and Marx.

On the level of methodology, he was a forerunner both in the area of mathematical economics and applied econometrics and in the area of using data and the concepts of accounting in economic analysis.

Without any doubt, Political Economy suffered from having neglected his methods during the Nineteenth Century.

After his time there was a tendency toward more abstraction and generality in spatial economics. His immediate successors only modernized his analysis without innovation [23][24][26][39]. Research was to become the work of theoreticians for the most part and, as a consequence, a gulf appeared between abstract and concrete analysis. Spatial empirical economics was developed through trial and error methods by researchers who lacked the proper tools to approach it.

---

(1) See paper by Schneider in Econometrica (January, 1934), trans. Anne Von Bibra Sutton under the title "Schneider on Thunen", in The Development of Economic Thought, ed.H.W.Spiegel, John Wiley & Sons, New York: Chapman & Hall, 1952, p.445ff.
(2) See Mathematical Appendix I,1.
(3) See Wolkoff, article in Journal des Economistes, 16, 1857, p.239ff and J.B.CLARK, "De l'influence de la terre sur le taux des salaires", Revue d'Economie Politique, 4, 1890, p.252ff.
(4) Cf. Mathematical Appendix I,2.
(5) Alfred Marshall, Principles (2nd ed.), p.556; he notes that Thunen was the first to use it in this context.
(6) Lorenz Von Stein, "Geschichte der sozialen Bewegung in Frankreich von 1789 bis auf unsere Tage", Bucherei fur Politik und Geschichte des Drei Masken Verlages, Munchen, 1921.

Chapter 3. From Thunen to Weber

After Thunen and his disciples, no new advances in the field of spatial economics were made until the end of the Nineteenth Century. During the second economic and technical revolution of capitalism, a transition was being made from an essentially agricultural economy, with animal-drawn means of transportation, to industrialization with its railroads, urbanization, concentration, etc. This change was felt especially in Germany, the main center of spatial thought.

At this time, location analysis followed two separate lines of thought which only today are becoming re-united. Some authors - mainly Anglo-Saxon - emphasized descriptions of location phenomena, which consisted of enumerations of location factors. German theoreticians, on the other hand, developed deductive analysis of spatial problems, particularly the location problem of the individual firm.

From Thunen until the nineteen-fifties, such abstract theory in spatial economics has been virtually a monopoly of the German school of thought [670]. The internationally dominant position of England, and later of the United States, seems to have blurred the vision of Anglo-Saxon scholars, relegating to the background local differences within the national economy and centering attention on the dynamic aspects of industrial development and international exchange. The reaction of the German school of thought to the classical teachings that precipitated the coming of the German Historical School prepared the ground for contributions within the domain of spatial economics (Raumwirtschaft).

Section 1. THE ANGLO-SAXON DESCRIPTIONS

As early as 1835, Ure enumerated the factors of location. He mentioned cheap energy, adequate population, existence of a port, warehouse or market place, and the introduction of innovations by industrial entrepreneurs [6].

It was not until the end of the Nineteenth Century and the beginning of the Twentieth that the problem of location was re-examined. American authors, such as Ross and Hall, proceeded to purely enumerative descriptions, like that of Ure, but they were more systematic [19][20][22]. The cumulative processes of concentration (agglomeration economics) were also perceived, but considered separately from other factors of location (1).

Hall was interested in the question of why a given site is specialized in a particular industry. If the costs of transportation are high, they determine location; if not, other factors affect it, for example, the nearness of raw materials or market, the availability of capital, climate, etc. Hall analyzes especially those industries where only one factor plays a dominant role.

Of all the orthodox Classical economists, only Marshall perceived some aspects of the spatial problem. In his "Principles" he points to the possibility of estimating in "monetary terms" the advantages of a location and the relationship between the cost of transportation and the distance to the market. In "Industry and Trade" he links his theory more effectively to location decisions by his analysis of increasing returns. He points to the attraction exerted by raw materials, climate, and markets of the product on the firm (2).

The German schools of "Standort" transfered the methods of pure theory to the area of spatial research and this line of thought was to lead directly to Weber's work.

Section 2. TOWARDS THE FIRST GERMAN DEDUCTIVE THEORIES

Weber's theory was undoubtedly influenced by the writings of Roscher and Schaffle, who tried to determine whether there were any natural laws in the spatial evolution of economic structures [9][13].

Roscher, still under the influence of Thunen, tried to develop an inductive theory and studied the natural factors of location. It seemed to him that the birth of an industry presupposes several conditions: a developed agriculture, a rich consumption, a population density heavy enough to allow for the division of labor, an abundant supply of capital and developed means of communications. In an economy where the division of labor is poorly developed, industry emerges near the centers of consumption which have themselves grown from large commercial centers. This is true in the case of industries that produce luxury goods or goods that are difficult to transport. If, on the other hand, division of labor has developed, industry will look for advantageous locations. These advantages may include raw materials, sources of power, capital, climate, etc. Roscher presents some examples of these types of location decisions and then from economic history discusses the influence of changes over time (e.g. the substitution of coal for wood). He also points out the advantages of large cities and discusses the differences between primitive and industrial countries, underlining the importance of trade between them.

Schaffle systematizes Roscher's analysis. He starts from the same basic hypothesis as Thunen does, with distance from the market being the essential factor in his model. Once again the concentric rings appear. Two conflicting tendencies also arise - decentralization and centralization. The latter is the stronger and is present in those industries that require capital, specialized workers, etc. Schaffle proceeds to classify industries and factors of production. His classifications foreshadow those of Palander and seem to have had a direct influence on them. Essentially, he distinguishes between those industries in which labor is the predominant factor whose immobility leads to immobility of location and those industries requiring several factors, so that their location is drawn to places where there are raw materials, capital, labor and energy sources. Schaffle studies the centralizing tendencies within each industry by looking at the existing relationships between the factors of production.

The German theory of spatial economics moved onto a deductive method of analysis with the works of Launhardt.

Full recognition was not given to Launhardt's work during his own life-time both because it was virtually unknown and because it was far too ahead of his time to be understood. As a specialist in transportation economics, it was to this field of research that he made his main contributions [11][12]. One of his works was rediscovered recently thanks to an early English translation [17][343] [622]. His contribution is surprisingly modern, in particular as regards his ideas on the location of communication routes, marginal cost pricing and the profitability of roads and railways, etc.

Transportation economics led Launhardt to take an interest in the theory of the location of production and the theory of market areas.

It is common practice among those who have commented on Launhardt's work to emphasise his transition from Thunen's theory of agricultural location to a theory of industrial location. By doing so they focus on a token change in the contemporary German economy. Of far greater importance is the fact that Launhardt narrows the scope by analysing the firm instead of an industrial sector. Launhardt shows how an optimum location point for the firm is determined by transportation costs as a function of the given locations of the production centers, raw materials and consumer markets [15]. He attributes transportation costs, which are proportional to weight and distance, to a system of mechanical forces, each pulling the firm towards given locations. Launhardt can then define the locus of the minimum transportation point. He builds it geometrically. Thus it is quite clear that Launhardt who wrote his theory in 1882 made his discovery before Weber who set forth his theory of the minimum transportation point in 1909. Launhardt's work remained unknown during Weber's life-time and thus History unjustly attributed this discovery to him rather than to his forerunner.

Elsewhere Launhardt tackles the problem of the size of market areas [16]. He studies the particular case of two sellers whose locations, at a certain distance from one another, are given. He goes on to study mathematically the laws of supply to those consumption areas. This analysis of the surface of market areas foreshadows much later contributions by Fetter, Palander, etc.

Weber, like Thunen, adopts a market reduced to one point and so also fails to see the close connection between the two problems. Englander was the first to recognize that these two problems are fundamentally one and the same, despite his inadequate synthesis.

---

(1) More recent, but belonging to the same state of thought, we can mention as examples Keir [38] and Black [53].
(2) Alfred Marshall, Principles of Economics, London, 1890, and Industry and Trade, London, 1919.

Chapter 4. Alfred Weber

INTRODUCTION. HISTORICAL AND THEORETICAL BACKGROUND OF THE WEBERIAN ANALYSIS

Alfred Weber was surprised to discover that, since Thunen, the problems of location had been neglected or abandoned to the geographers, despite the economic, social, and technical importance of these problems in an era of international and intranational migrations [27] [36]. Weber restricted his interests to industrial location, since Thunen had already constructed a theory of agricultural location. He does not, however, try to explain the location of trade, credit and capital. The distinction, which has gradually been disappearing, between agricultural and industrial location theories was significant in spatial economics up to Losch, and even afterwards.

Weber was aware of the growing gap between Thunen's model and the economy of Thunen's time on the one hand and contemporary economic thought on the other.

Thunen wrote in an era of animal-drawn means of transportation and rural economy, and Weber in one of steam, railroads and industrialization.

Weberian theory is a projection of pure economics into the spatial domain, elaborating laws which are abstract, mechanical, independent of any economic system (Wirtschaftsart), and applicable in the modern world. In this way, it was a reaction against the prevailing historical school of thought (German "Historicism"). Established political economy was divided into three sectors: production, distribution, and consumption. While Weber believed in limiting his study to the integration of spatial factors into the first sector, his desire for a total explanation which would account for all interactions led him to enlarge the scope of his analysis. Therefore, he made allocation of consumption dependent on the allocation of production and distribution (1), so that the location of production cannot be analyzed without analyzing consumption.

Weber's ambition was to bind to a deductive theory of location an inductive theory of those forces which the deductive theory leaves unexplained. In "Grundriss...", he used data for the German economy since 1860. But this application to the spatial domain of the method of successive approximations did not succeed here, nor has it elsewhere, to explain actual location patterns and shifts.

Only on the level of abstract theory does Weber's work constitute a significant contribution. The inductive part of his work, which was carried out mainly by his followers, does not succeed in demonstrating the applicability of his theoretical model to historical trends (2).

Section 1. PARAMETERS AND ASSUMPTIONS

Weber's process of abstracting starts by isolating and grouping factors of location. The "locational force", i.e. the forces which operate as economic causes of location, is defined as a cost advantage, and the locational unit, as the productive and distributive processes of a given product.

Location factors can be of a general nature (transportation, rent, labor) when they are present in all industries, or of a specific nature (perishability, humidity, etc.) when they apply only to specific industries. They can distribute activities regionally (transportation, labor) or they can concentrate or disperse them within a region (agglomeration or deglomeration factors) by the interactions within and among industries. There is also a distinction between natural and technical factors on the one hand and social and cultural factors on the other. While the former are capable of measurement and incorporation into pure theory, the latter are not.

Weber outlines four steps in the process of production and distribution: acquisition of the location and the equipment; procurement of raw materials and energy; organization of the process itself; and delivery of the product. He sums the costs of these four steps, plus the general expenses. Of these costs, seven are relevant in his analysis: profits, interest on capital, rate of amortization of fixed capital, cost of raw materials and energy, wages, cost of transport and general expenses. Weber eliminates profits, since they do not constitute an element of price determination but are a consequence of price. Profits are incorporated into the costs only during the final steps of the process and do not become a factor of location unless we go beyond the pure Weberian theory. General expenses are also eliminated either because they are artificial factors (taxes, etc.) or because they are not significant for location. All the other elements arise from pure theory. The distinction between the regional and non-regional factors of location is very important. The rate of depreciation of fixed capital and the interest rate, in Weber's theory, do not vary regionally. The cost of real capital, fixed and in circulation, depends on the early steps of production and does not bring in any new element dependent on it. On the other hand, wage and transport costs appear as regional factors. All the non-regional factors are agglomerative or deglomerative. The price of land varies as a function of local, not regional, agglomeration, except for agriculture for which Weber relies on Thunen's system.

Weber determines the optimal location based on transportation costs, and then introduces distortions caused by labor and by agglomeration and deglomeration forces.

The selection of factors of location imposes first a series of limitations on the generality of the Weberian theory. To this must be added a certain number of simplifying assumptions. Weber therefore assumes as given the locations of raw materials and places of consumption, the immobility and unlimited supply of labor at fixed wages, and transportability of raw materials.

Section 2. THE THREE TYPES OF LOCATION ORIENTATION

1. The Point of Minimum Transport Cost

Like Launhardt, Weber attributes the costs of transportation to the weight of the product and the distance. Complicating elements are arbitrarily simplified. A uniform transport system replaces the variety of systems existing in reality. Differences in rates, including those for less-than-carload shipments, are converted into variations of weight or distance. Real weights are therefore replaced by "ideal weights", which standardize weights and distance. The notion of "ideal weight" standardizes for the differences inherent in the nature of the region, in the ways of transportation, or in the nature of the goods transported.

The laws of orientation as a function of transport costs (Transportorientierung) are found by constructing a "locational figure" (Standortsfigur) drawn by linking with straight lines the points that constitute the place of consumption and the centers from which raw materials and energy are obtained. Weber draws in this way a geometric figure that serves as a framework for the location problem. For example, two sources of raw materials and one place of consumption give us a triangle. All that must be found is a point of minimum transport cost within this figure where the corners represent poles of attraction of variable strength similar to a Varignon frame (3). The raw materials and the finished product move along lines linking this point of minimum transport costs to the corners of the figure. The location of this point is determined by the force with which the corners attract the location in their direction. This force is a function of the "ton-miles" that can be saved by approaching the corresponding corner. Ubiquitous raw materials affect the final result to the extent that they increase only the weight of the final product and, in such a manner, strengthen the force of attraction of the place of consumption. Inversely, Weber distinguishes between "pure" and "weight losing" raw materials. The latter lose part of their weight during the production process and thus strengthen the attractive pull of their source locations.

The combination of all these elements gives the "locational weight" (Standortsgewicht) of the firm under consideration. The ratio:
number of units of weight of localized raw materials / number of units of weight of the finished product, determines the orientation of the production point. Weber calls this ratio the "material index" (Materialindex); it measures the total unit weight. Depending on whether the index is larger or smaller than one, the most powerful attraction will be that of the raw materials or that of the market place. But in practice, this theoretical index is replaced by the ratio:
units of weight lost by the localized materials / units of weight of the ubiquitous materials = Freight Index.

Once all variations are reduced to differences in distance and weight, various possible freight systems and the interaction of multiple combinations of transport lose their special significance. They distort only the forces of attraction by altering the extension and the shape of the "locational figures". In the same manner, differences in the prices of raw materials are expressed in terms of transport.

Looking at an entire industry, Weber adapts his hypothesis of reasoning, without changing his level of observation, by drawing new "locational figures". If only one raw material is used, or if several pure or ubiquitous materials are used, usually those deposits of raw materials which are closest to the place of consumption will be used. Frequently, a center of consumption is served by several centers of production using, according to the freight indices, one or another deposit of raw materials. If one source is found to be insufficient, others less well situated will be used and production started there. Several "locational figures" with different freight indices are used. Conversely a large deposit can serve several centers of consumption, if alternative deposits would generate greater transport costs.

In long term development, location at the sources of localized weight losing raw materials is favored. This, according to Weber, is because technical progress tends to increase the weight loss, thereby increasing the freight index of shipment. The growing concentration of population means that there is a risk of exhausting the ubiquitous materials and thus weakening the force of attraction of the place of consumption.

2. Labor Orientation

Weber defines labor orientation as the deviation of location from the point of minimum transport cost to a point of lower labor costs. In this analysis, he considers only the variations in the level of local wages due to the subjective efficiency of labor and disregards objective efficiency. Objective efficiency is a function of the degree of capitalization and is revealed by an analysis of the firm's dimension. Subjective efficiency is a function of a particular geographic distribution of the population and appears to Weber to be the only cause of regional variations in the cost of labor. Weber maintains that the differences in the cost of labor are more important than just those of the nominal wage, which may not reflect differences in efficiency. Finally, he retains in his analysis only wage differentials between points, the study of the contemporary German economy having shown him that disparities between regions are only meaningful on the average and have no relevance for the individual entrepreneur.

Deviations from the point of minimum transport cost can occur in any direction. Therefore, around this point, a set of new points can be determined for which additional transportation costs are equal. Weber calls these geometric curves isodapanes and he distinguishes one called a "critical isodapane" for which the saving in labor costs is equal to the extra cost of increased transportation to the new location, per ton of product. The location of the firm will deviate toward the point of lower labor costs if this point is located inside the area delimited by the critical isodapane: i.e. if it is located on a lower cost isodapane (4).

For any firm there are three decisive elements affecting location: the geographic position of the "locational figure" and of the labor location; the pattern of the isodapanes around the point of minimum transport cost; and the index of labor savings per unit weight of product. From these elements originate the laws of attraction of labor. The locations of the "figure" and of labor are given. The pattern of the isodapanes depends on the transport rates, "material index", and "locational weight" for the individual industry under consideration. Finally, the index of labor savings depends on the percentage of reduction possible in the labor costs per ton and upon the absolute level of this reduction. This Weber calls the "index of labor cost". The formula that gives us the orientation of the firm as a function of the labor location is obtained by relating the factors acting on the isodapanes to those affecting the index of labor saving.

The higher the "index of labor cost", the stronger is the attraction of labor; and the lower the "locational weight", the greater is the deviation of the firm, because of the stretching of the isodapanes. According to the "material index", the minimum point of transport is located in the interior of the "locational figure" or at a corner and, consequently, deviation will be more likely in certain directions than in others. For large distances, the "locational figure" is reduced to a point and the "material index" loses its significance.

Finally, the "index of labor cost" divided by the "locational weight" is a measure of deviation, in Weberian terminology, referred to as the "coefficient of labor". When the denominator is reduced to unity, labor costs are expressed per ton and the coefficient becomes comparable for all industries.

At the level of an entire industry, if the individual units deviate easily, the labor locations will influence all the "locational figures" around them. High "coefficients of labor" will lead, thus, to a concentration of industries; low coefficients to their dispersion. Concentration can result in closing down remote deposits of raw materials and opening closer sources. The deviation then provides a transportation saving in addition to labor savings.

Both labor and transportation orientation are dominated by two environmental conditions: the density of population and the stage of technological development. A low density of population means a lengthening of the average distance between "locational figures" and labor locations. For populations with little variation in level of development, a low population density implies, in addition, regional homogeneity of productivity, wages, etc. It will be more difficult, in this way, for the deviational attraction of labor to prevail than in the case of a dense population. Finally, if the cost of transportation declines with distance, then the isodapanes are spread and the labor locations influence the individual units more strongly and cause concentrations.

The long-run tendencies of development re-enforce the attraction of labor locations due to the decrease in transportation costs and the increase in the density of population. However, mechanization, by reducing the quantity of labor per unit of "locational weight", favors an orientation towards the point of least transportation cost. This convergence of the isodapanes and the decrease in possible labor reduction draw the "critical isodapane" and the point of minimum transport cost closer together. They constitute a force which is opposed to the attraction of labor. The amount of spread of the isodapanes and the position of the "critical isodapane" with respect to the point of minimum transportation cost are modified in a decisive way depending on whether the increase in cost of the raw materials utilized is greater or less than the reduction in the cost of transportation.

3. Agglomeration

The third set of factors influencing location, as discussed by Weber, is the economies that result from "agglomeration" (a concentration of firms). The rise in land rent caused by concentration creates an opposite force of "deglomeration", but Weber sees this as a negative force of agglomeration. He studies the resultant of these two forces, while assuming that the "agglomeration function" is parabolic and the "deglomeration function" is linear (5).

The result is an index of saving in unit cost which depends on agglomeration, defined as the advantage, or cheapening, of production or of marketing resulting from concentration. Various individual indices of saving per unit of product may correspond to each stage of concentration and give us, using Weberian terminology, a "function of agglomeration economies". On the other hand, a fixed index can be the result of a definite amount of concentration. The advantages of concentration (compression of general costs, better adaptation to market conditions, etc.) lead either to an increase of the size of the firm, or to local integration of several firms.

Weber defines the laws of agglomeration by relating the savings of agglomeration to the rise in costs caused by necessary deviations, expressed in additional costs of transportation and/or labor.

With respect to the point of minimum transportation cost, and in the case of a fixed index of agglomeration economies, the isodapane method permits comparison of the savings in cost obtained by concentration and the rise in the cost of transportation caused by the deviation. New "critical isodapanes", for which these two magnitudes are in equilibrium, will thus appear around the units considered. Agglomeration will occur if several "critical isodapanes" intersect, and if the sum of the quantities of production of each individual unit which participates in the same overlapping segment reaches the quantity required to be able to benefit from the economies of agglomeration. Concentration will then appear in the common segments of the "critical isodapanes", at a new point of optimal transportation cost for the larger establishments. This new point is determined by a new "locational figure" which unites the center of agglomeration to the sources of raw materials and the markets. Between the various points of possible agglomeration, the point chosen by each firm will be the least distant from the former minimum point of that firm. The closer the other firms the less distant will the new location be, so that the common segments of the critical isodapanes will involve larger surfaces. The result of agglomeration may be to open up new deposits of raw materials which are now closer, to the disadvantage of the old deposits, thus increasing the attraction of the agglomeration center by reducing transport costs.

Given the assumption of a "function of agglomeration economies", the problem remains of determining which unit of agglomeration will prevail. Since the tendency toward concentration is the simultaneous result of various indices, several critical isodapanes will surround the points of production. Each of these critical isodapanes corresponds to a particular index of economies for a particular industrial complex. The critical isodapanes of the largest concentrations are the farthest away from the firm's minimum transport cost point and can intersect the greatest number of other critical isodapanes. A larger total production is, however, necessary to justify these larger concentrations.

The laws of agglomeration depend on the scale of the critical isodapanes (which determine the rate of increase of agglomeration economies), on the distance between the common segments (on which depends the choice of the agglomeration), on the distance between points of production (which determines the possible intersections of the critical isodapanes), and, finally, on the quantity produced by each unit. The last two conditions are combined into the Weberian notion of "density of industry" which depends on the general conditions of the environment: the density of the population (which fixes the quantity of production of the industrial units), and the number of population centers (which distributes these units). The distance between the critical isodapanes depends on the "locational weight" and the cost of transportation. Thus, the deviation of the agglomeration depends, on the one hand, on the two conditions inherent in the character of the industry - the economies function and locational weight - and on the other hand, on the two exogenous conditions - the cost of transportation and the density of the industry.

The attraction exerted by agglomeration advantages can bring about a deviation with respect to the locations of the units of production at the labor centers. Weber points out that such locations often give rise to an "accidental" agglomeration, bringing about a saving in costs. In order to draw the isodapanes, the labor economies and the accidental agglomeration economies must be compared to the economies of pure agglomeration. While accidental economies may rest solely on three of the factors of pure agglomeration (locational weight, cost of transportation, and density of population), the fourth (the economies function) should be particularly strong if a pure agglomeration is to occur.

Failing to obtain deductively the economies function, Weber's general idea is that those industries whose products derive a large portion of their value from the production process (value added or Formwert) are the only ones to present a sufficient percentage of compressible costs to obtain benefit from pure agglomeration. The "index of manufacture" expresses the value added per ton. Its two components are, in different proportions, labor and capital. These possess an unequal reinforcing power on the deviational force of agglomeration, since the use of capital implies an increasing "material index", while labor is a pure factor of agglomeration. A new index, the "locational ton", relates the index of manufacture to the total weight which has to be transported, and the value added per "locational ton" gives a "coefficient of manufacture" (Formkoeffizient). Industries will show a tendency to agglomerate in direct function to their "coefficient of manufacture".

Thus, the agglomerative force can influence industries already oriented either by labor or transportation. In the first case, however, if participation of labor in the integrated process is less important than the participation of capital, the "coefficient of manufacture" will be low in comparison to the high "locational weight". Conversely, if the participation of labor is greater than that of capital, the attraction of labor location will give rise to a technical agglomeration which will reinforce the power of attraction of the labor center.

Tendencies in industrial development which modify the conditions of the concentration, e.g. density of population, cost of transportation, and the "coefficient of manufacture", favor its deviational attraction because the demographic density is increasing and the cost of transportation is falling. The impact of the "coefficient of manufacture" is less clear, since it is weakened by the "locational weight".

4. Total Orientation

The process of production can take place successively in several locations. The firm includes several establishments. Weber, in his theory of "total orientation", shows how the location pattern of these establishments obeys the three fundamental laws of orientation. The reasoning is substantially the same as for the individual firm. The division of the process modifies nothing but the form and dimension of the geometric figures or the size of the various indices and coefficients defined by Weber.

Similarly, Weber takes into account the fact that the processes are not independent of each other, but their independence does not modify the laws that have been developed.

Section 3. THE SYSTEM OF LOCATIONS

At the end of his book, Weber poses the question of how possible groupings of production units are distributed spatially within the geographical limits of a country.

A basic merit of Weber's work is thus his integration of a theory of systems of location into his theory of location. He sees clearly the implications of enlarging his scope of observation. He makes the transition from individual to system theory better than he makes the transition from the analysis of the firm to that of the industry. The four assumptions, from which his theory of location was developed, are now determined by existing spatial patterns. Thus, the location of the places of consumption, raw materials, labor, and the supply of labor become variables in the analysis dependent on actual spatial conditions.

In an attempt to discover which is the dominant force integrating the various parts of a spatially isolated economic system, Weber imagined the establishment of a new system in an empty zone. Several layers of spatial distribution develop. The first layer, agriculture, is distributed in accordance with Thunen's model. A portion of the population is distributed over as large an area suitable for agriculture as is necessary for the production of agricultural products. This layer supports agriculturally related industry which constitutes the first industrial layer. The places of consumption and the sources of raw material for this first industrial layer are given by the spatial distribution of the agricultural layer. This industrial layer depends on the "locational figures" of the first (agricultural) layer. The spatial distribution of economic activities is thus structured by four groups: the industrial population engaged in the first industrial layer; the population engaged in the distribution and trade of the goods produced; pure consumers; and those who satisfy the needs of the previous layers.

The primary industrial layer determines a consumption area and thus creates the foundation of the location pattern. The division of labor causes the formation of numerous "substrata" which are superimposed on each other, interrelated, and of decreasing size.

A secondary industrial layer is articulated, based on the primary one, and together with the agricultural layer and the first industrial layer, orients the whole system. The population which engages in distribution and those who only consume strengthen the different parts of the system proportionally. The location of the consuming classes (stockholders, shopkeepers, etc.) which is the "central organizing layer" seems to Weber to be the most aberrant. He introduces finally a fifth layer, which depends on the central organizing layer, including an industrial part and an organizing one, which is both central and local and has interdependent "substrata". The local organizing layer is embraced in the first three layers - the agricultural and the primary and secondary industrial - and serves as a strengthening element.

The locational forces which connect the different layers interact, in a systematic way, while Thunen's central city remains, according to

Weber, the "unexplained" phenomenon of the distribution of agricultural locations. But, in the Weberian construction, the foundation remains the agricultural base upon which the non-agricultural activities are then located. Weber recognizes that the connection between the first industrial layer and the agricultural layer is difficult to explain, since the rural population, scattered as it is, can be supplied only by the industrial agglomerations of cities. But he asserts, without proving it, that rules identical to those of the higher layers, where the population is more clearly distributed, apply also to these first layers.

In this manner, Weber tries to show how his theoretical assumptions can be replaced by real conditions. With respect to the unlimited supply of labor, a basic assumption of his abstract theory, Weber leaves this problem to be solved by a "realistic" theory, feeling that it depends on the economic system in the meaning of Sombart.

Aside from the imperfections and the simplified character of his model, Weber made substantial advances, including anticipating in his model Losch's theory of market areas and networks.

CONCLUSION

Weberian theory considers a closed economic system. However, Weber did not ignore the international aspects of the described phenomena [32]. He found fault with the theory of international division of labor and suggested the application to the field of international trade of the analysis of the forces that explain industrial location.

Instead of starting from the international division of labor he thought it proper to start from the "natural forces" which manifested themselves in the distribution of economic forces, that is: the distribution of the agricultural population around the historic centers of culture and population in accordance with Thunen's rings; the orientation of industry as a function of the cost of transportation when it does not use weight-losing raw materials; the attraction of industry which employs such impure raw materials to the deposits of these materials and sources of energy; and the concentration, as a function of the location of labor, of industries which have a high labor cost per ton. Thus, with this Weberian criticism, the classic theory appears limited to the last of these cases, which moreover corresponded to the prevailing economic situation in England at that time.

Nevertheless, the most important historical contribution of Weber is still his mechanical theory of location (Standort) for a closed system. The Weberian school of location was to survive up to modern times, remaining contemporary with the most modern theories, but continuing to belong to a stage of spatial economic thought that Predohl, after trials and errors which were sometimes interesting, attempted to surpass.

(1) Except for those who are not consumers, for example the stockholders.
(2) See [27] Part.II: Die deutsche Industrie seit 1860.
(3) Cf.Mathematical Appendix II,1. As early as the 17th Century, two mathematicians, Fermat and Cavalieri and a physicist, Torricelli had examined the problem of determining a point at which the sum of the distances to the corners of a triangle is minimal.
In the case of n points, the "Torricelli point" is the point, if it exists, such that the resultant of the unitary vectors aiming from that point to the n given points is zero.
(4) Cf.Mathematical Appendix II,2.
(5) Cf.Mathematical Appendix II,3.

Chapter 5. From Weber to Palander

Shortly after the publication of Weber's work, Bortkiewicz and Schumpeter began emphasizing the need for a general equilibrium analysis to integrate the partial theories of location (1). Furlan, in his article, recognizes fully the complicated interrelations between the numerous economic factors of location and "the spatial transformation of goods", but his line of development follows very simplified market models [34].

It was the influence of the Lausanne School on German political economy that was to give major impetus to Predohl's effort. This influence made inevitable the works which combine spatial analysis with general equilibrium theory.

Section 1. Andreas PREDOHL

The essential contribution of micro-economic marginalism to economics can be defined as a perfect systematization of the phenomenon of substitution. From Walras to Hicks, the substitution effect is at the heart of economic theory as defined by the marginalism. It is defined in rigorous mathematical terms and applied to the system of equilibrium of both producers and consumers. The introduction of time by modifying the assumptions necessary to marginal analysis, undercuts the basis of the static system of the marginalists. Similarly the introduction of space, by Predohl, undercuts the basis of a nonspatial system, but at the same time enlarges the possibilities for the utilization of the concept of substitution already elaborated by traditional doctrine [51][60][63].

Predohl built the first bridge between marginalism and the theory of location. His work, contemporary to the classical models of Walras and Cassel, is also related to the partial theories of location developed by his two principal predecessors, Thunen and Weber.

Predohl subjects these diverse theories to a critical examination and points out their limitations. Above all, it appears that the theories had not been satisfactorily interrelated; there was a gap between general economic theory and the special theories of agricultural and industrial location. The two areas of research had been explored separately so that the resulting analyses were lacking in homogeneity and rigorous logical connections. The proper role of a general location theory would then be to fill this void by the application of scientific method, not merely empiricism, in order to derive special location theories from general economic theory. Predohl tries to show how an analysis of location could be deduced from an abstract model of the formation of prices. Using the concept of substitution to introduce space into traditional Walras-Cassel models, Predohl develops a general and abstract theory of location. In this way, he is able to relate and compare his theory to the theories of Thunen and Weber. Although he distinguishes in their works the obvious differences between their agricultural and industrial aspects, he fails to emphasize the differences in their scales of observation.

Despite its ambitious scope, Predohl's theory contains its own limitations. It is limited to the level of the firm and to the very short run, "with all other factors remaining constant".

While Predohl attempts to approach reality by making his model more and more specific, he does not prove that this procedure will lead to realistic conclusions.

1. Substitution and Location

Having posed the basic location problem, Predohl concentrates his attention on the results of change in the location of the production process, while output remains constant.

The totality of these effects can be reduced to the changes in the input composition of the final product. Here the author introduces the principle of substitution borrowed from Marshall and Cassel. This seemed to be the most practical solution to the economic problem of factor combinations, and thus of location. Predohl shows that production and location are one and the same problem. Location becomes a simple variation of the general economic problem as it was conceived at that time. He feels, therefore, that he is correct to apply to the problem of location the same theoretical approach as that applied by other authors to the non-spatial theory of the firm.

Predohl attributed all changes in the location of a firm to the substitution of the various factors of production located at various sites. This substitution is a function of the relative prices of factors and their transportation costs. These factors appear to be affected by a coefficient of local weighting and the variations appearing in these coefficients bring about modifications in the combination of factors, and consequently changes in location. Although in classical theory the equilibrium system is devoid of any spatial implications and substitution takes place at a single point, in Predohl's scheme, substitution implies a change in location. Thus, spatial equilibrium is introduced.

The need to divide the infinity of differences affecting the factors of production into three categories led Predohl to invent obscure concepts - somewhat in the manner of Weber.

Following a change in location, the changes taking place in the quality and quantity of land are translated in Predohl's terminology, into changes in the number of "land use units". The changes in the quality and quantity of capital and labor are likewise expressed by changes in the number of "capital and labor use units". Finally, the differences in the quantities and qualities of transportation caused by the variations in distances using different systems are conceptualized as the number of "transport use units".

If the total land use units are given and if the other expenditures are all grouped together, the substitution of the land use units by other factors, or conversely, following a change in location can be measured. Other substitution effects can be measured in the same manner. If one or more raw materials enter into the production process, Predohl suggests the division of the transport input group into sub-groups. Where no transported raw materials are employed, it is only necessary to take into consideration the transportation of the finished product from the place of production to the place of consumption.

The location problem is thus reduced to a system of points of indifference, each pair of groups of interchangeable use units being subordinated to a larger group. The point of indifference (the point of minimum cost) fixes the relative rate of substitution of the factors taken two by two.

In his attempt to develop a general theory, Predohl, like Weber, is troubled by the question of the location of consumption. He subordinates this problem to the problem of location of production. He believes his theory to be limited by only that minimum part of consumption, mainly of owners and stockholders, whose location is determined independently.

After insisting on the logical strength that his analysis derived from its deductive character, Predohl remarks that substitution analysis does not show which factors are interchangeable and why they are substitutable. The limitations of the Weberian mode of reasoning, in technical terms, had not yet been overcome by spatial analysts. The principle of substitution seemed still more formal, even to Predohl, so that he did not anticipate the possibility of obtaining any supplementary results without using some other method of reasoning.

In any case, questions of factor substitution can only be answered in terms of specific cases. Predohl attempted to arrive at certain empirical generalizations by considering only the costs of those factors which are substitutable. For example, in Predohl's time, the steel industry was traditionally located near coal mines and not at the sources of iron ore. The textile industry was located near labor centers. Predohl thought that it would be possible to generalize further when only technical processes were involved. The rate of substitution could then be expressed in technical and quantitative terms. The weight of the coal to be transported to the steel mills could be compared to the weight of the iron ore. The author thus anticipates a supplementary theory of the technical process of substitution. However this "supplementary" nature is essential: prices must be both given and constant, if not they are determining factors in location.

In the tradition of neo-classic theory, Predohl's attempt at a general theory of location is through the synthesis of specific cases. His theory is an attempt to make an abstract generalization starting from partial analyses of location. Hence, it is useful to understand how the author relates the framework of the general theory to the partial theories of location by manipulating the linking mechanism between these theories.

2. The General Theory of Location

Predohl's theory is deduced from the Walrasian theory and attempts to combine the theories of Thunen and Weber. Apparently unaware of the differences in the scales of their observations, Predohl emphasizes the nature of the surfaces they introduced as the fundamental distinction between their theories. If there is a continuous plane, as Thunen postulates, the factors of location are the location rent and the cost of transportation. If there is a discontinuous surface, as Weber postulates, with centers of consumption, labor, and raw materials dispersed in it, the factors of location include the diverse local costs of capital and labor. Nevertheless, with the factors of production selected differently, Thunen and Weber study nothing but combinations. The two analyses are implicitly conducted in terms of substitution and their limits are related only to those factors of location that are selectively retained, on the basis of their relevance to the theory.

Thus they appear to Predohl as particular cases of his general formula.

A specific theory would not be deducible from Predohl's general theory if it considered only technical relationships. Weber especially ignores the basic importance of prices as determining factors when he reduces the cost of transport to weight and distance. Only when the prices are given and constant is it possible to work solely within the framework of technical substitution. The theories of Predohl and Weber coincide only under this condition.

In conclusion, Predohl's theory was for a long time considered authoritative in the field, despite the controversies that it aroused at the time of its appearance [55][59] (2) but it still has severe limitations. Because it rests on the principle of substitution, it functions only at the level of the individual firm and in the very short run. It continues to depend on the twin assumptions of the fixity of technology and the continuity of space. These permit continuous substitution to take place.

Nevertheless, Predohl's theory served to sharpen the focus of abstract analysis which led to Isard's work on the spatial equilibrium of the firm.

Above all, the problem of general spatial equilibrium had been posed. While traditional analysis presented economic relationships as a non-spatial system by means of mathematical equations, the question remained as to whether or not such a system of equations for a spatial economy could be constructed and solved. Despite the difficulties which this question raised, this central preoccupation was to lead to the basic works of Palander and Losch. Taking account of the contemporary evolution of the neoclassic view, especially the work of Chamberlin, these theories were to be constructed with the assumption of limited competition. Finally, Anglo-Saxon contributions begin to acquire an importance, at times decisive, in the evolution of spatial economics which up to then had been almost exclusively the domain of authors of the German school.

Although Predohl introduced the methods of microeconomic analysis into location theory, the influence of Weber was still strong enough to generate a whole tradition of spatial research. His theory of stratified layers induced other authors to attempt a total approach to the problem of the interdependence of locations.

However, the impact of the neoclassical theory of prices on spatial economics was to induce another line of thought which, by treating the spatial phenomena of the market in modern terms, was to lead directly to Palander.

Section 2. FROM ENGLANDER TO CHRISTALLER

Even though contemporary with the work of Predohl, the contributions of Englander [43][54][55] and Ritschl [61] do not belong to the same school of thought. They try to analyze the general equilibrium of a spatial model, but their contributions are still linked to Weber's theory of stratified layers.

For Englander, the pure theory of location is part of a general analysis of the "local conditionality" at the heart of an economic system. Every entrepreneur, in choosing his own location, foresees the various prices of inputs attached to the various sources of supply and the prices that he can expect for his output at various markets. Once a site is chosen he himself then influences the prices. By considering these interrelationships, the theory of "local conditionality" determines simultaneously, within the model, both differences in local prices and the location of economic activities.

Englander classifies the factors of production in several groups: mobile or immobile, ubiquitous or localized. He conceives of immobile factors as entering production with an infinite weight.

His pure general theory includes specific theories of agricultural and industrial location as interrelated special cases and not as distinct components.

His general method of analysis follows a direction only slightly different from that pointed out by Weber. Englander studies the spatial model of primary production; that is, a land and forest economy in which all households are self-sufficient. Then specialized products are cultivated on land of a particular quality and the consecutive changes in the spatial structure of the economy can be observed. Agricultural industries, mines, manufacturing, and other real world phenomena are introduced and the consequent spatial realignment of economic activities is noted.

Ritschl admits that, historically, location decisions are made relative to existing location patterns. He takes Weber's classifications and describes the development of the activity layers during the periods of the village, the city, the territory, the nation, and the world.

Hawtrey takes the essentials of Englander's scheme, but emphasizes the importance of commercial centers and money markets [49]. In addition, the consideration of fixed costs leads him away from the traditional assumption of transportation costs as proportional to weight and distance.

Meanwhile, starting from the location theories of Englander and Predohl, Weigmann, whose contributions remained unknown until recently, tries to develop a general model of spatial economics [57][73][78][90]. Using a complicated style and difficult basic concepts, he looks for the foundations of a "realistic" theory, including the totality of the spatial structure of the economic process, the extent and the spatial links of the markets, and the spatial interrelations of economic quantities.

In conformity with the contemporary tendency in spatial thought, Weigmann's analysis relates the theory of spatial economics to that of monopolistic competition. The markets, no longer viewed as points but as surfaces, are limited in space because the mobility of the factors and products is limited by diverse obstacles. The competition among products, and among the factors themselves, in different places is imperfect. The principle of perfect competition, still generally accepted in Weigmann's time, appears to him to be inappropriate for spatial analysis. The existence of a physical space alone implies a limited competition.

From the methodological point of view, Weigmann studies the economy as a whole in its vast organization of spatial markets. He looks for a "realistic" and functional matrix of the configuration of the economic life within which the diverse elements are weighted according to their importance. Following this method, he tries to determine the "basic configuration" (Grundgestalt) of economic phenomena. With the help of this, he would be able to master and to order systematically the whole of the innumerable spatial forms associated with economic processes.

Aiming for a realistic description of the space-economy, Weigmann must complicate his problem by introducing the time element. He does this by assigning temporal coordinates to his markets and processes.

For him, the dynamic problem becomes one of choosing the period of time which provides within the spatial distribution of the markets an area of competition valid as a "basic configuration". The difficult concept of "relative maximum" is thus introduced. As the amount of physical space that must be traversed increases and the spatial resistance to be overcome by an economic object in movement rises, the period of time required for such a movement increases until it reaches a maximum. Beyond this point further spatial movement is unlikely. When the period of time reaches its maximum, the field of competition is bounded and further competition ends, since the force of competition is not felt beyond this boundary. This principle of Weigmann's defines the "basic configuration" as the unit of space related to the largest relative time period, having the greatest stability and relative permanence.

The need remained to localize the "basic configuration", to represent in a formal way the multitude of interrelated individual markets, the layers of markets, and their densities.

Weigmann starts by classifying markets according to their structure. Every market for an individual product possesses a particular structure offering a certain resistance to change. It is oriented in a specific fashion by the factors of attraction: labor, capital, and land. Some structures are active and change frequently, while others are passive and change slowly. The latter are of a relatively permanent nature, persistently inactive, and are essentially the "basic configurations". Their combined structures determine the fundamental structure of the space economy. The changing markets are considered as accidental or secondary. Their movements are characterized as minor modifications, conditioned, up to a point, by the "basic configuration" already determined by the nucleus of the more stable markets. A fundamental change in the spatial configuration of the economic model implies a change in the nucleus of relatively immutable markets.

Weigmann makes his analysis more precise by referring to the specific structures of the markets for the factors of production: land, labor and capital. The markets for land and labor constitute the first layer of the "basic configuration". The product market is oriented around this configuration either directly or indirectly through the movement of semi-finished goods and intermediate stages of production.

The land market is a spatially continuous land area, even though each plot of land is distinct and immobile and consequently its particular market lacks a spatial extension. Weigmann imagines a space economy already in existence and exerting a hypothetical total demand which in turn defines the boundaries of the land market, the peripheral area being considered marginal land. From the spatial point of view, the supply of land is not perfectly inelastic, as the land market can be modified by additions or substractions on the boundaries of the land under cultivation or by the changes in the intensity of cultivation, by the methods of exploitation and organization of each individual unit of land.

Weigmann relates the size and the nature of the hypothetical demand to the labor market. As compared to the land market, the labor market is less rigid. The forms of immobility and inelasticity of labor are numerous. Weigmann starts by considering one form of mobility: migration. His theory divides the labor market according to different periods of migration (seasonal, cyclical, and secular). A long run process displaces labor from the farms or rural communities to the towns, cities and large metropolitan centers. This gradual structural movement constitutes one of the essential dynamic aspects of contemporary spatial political economy.

With respect to capital markets, Weigmann limits himself to making suggestions for possible investigation, distinguishing between real capital and financial capital. The former is less mobile than the latter. Capital goods, which in the wider sense of Weigmann's formulation, comprise all merchandise, should be classified according to their importance in production. At one extreme, capital goods would be permanently tied to a given production process. At the other extreme, capital goods would enter into "free combination". The markets of the different categories of capital goods and their spatial elasticities vary as a result.

Weigmann's analysis appears more comprehensive when compared to the traditional theory of location which generally regarded capital as the factor of "free combination" in the long run, disregarding the suitability of the existing equipment in the short run. His analysis tends to consider the mobility of a given combination of productive factors as a whole instead of its constituent parts.

Weigmann's economic model appears as a set of spatial relations affected by a rhythmic movement and with a nucleus composed of markets for land, labor, and capital, with numerous other markets superimposed, overlapping and crossing each other in an irregular fashion, and extending themselves at times over other systems.

It was at this point in 1933 that Christaller again poses, in less general but more rigorous terms, the problem of the structure of the spatial economic model, in his famous book on central places (Die zentralen Orte) in South Germany [76].

Christaller, as a geographer, looks for the laws which explain settlement in rural and urban spaces, or more specifically, the relationships between the number, size and spacing of towns. He followed the controversies which raged among German geographers over the relationship between the size of towns and their distribution in space: it was their differences in opinion over the concepts of the nineteenth Century pioneer Kohl which was to divide those geographers [8]. Christaller also followed the discussions amongst German economists such as Schaffle and Weigmann, synthesized by Schmidt [52] during which they analysed the influence of the spatial factor.

Christaller adopts a deductive method to develop a geography of settlements (Siedlungsgeographie). He opens his work by giving his theoretical foundations which are essentially economic. He then verifies the theoretical model by applying it to the concrete case of South Germany. The author presents his paper as a theory of the location of retail outlets and urban institutions to be placed alongside Thunen's theory of agricultural location and Weber's and Englander's theories of industrial location.

Christaller defines the town as the center of a regional community and the place which acts as the intermediary for that community's commercial activities; as such the town is the community's central place.

The central places which are different in size are ranked in order of size. The central places ranked the highest in terms of size control wider regions than the central places ranked lower; they exercise more central functions and possess greater centrality. However, for all central places, the sum of the distances which separate them from the surrounding rural residents, distances which those residents have to cover in order to procure "central goods and services" is minimal. It should be noted that higher ranking goods are supplied by the higher ranking central places and that lower ranking goods are supplied by both higher and lower ranking central places.

Christaller calls any area dominated by a central place a "complementary region" (Erganzungsgebiet). It is the locus for the two-way relationships between the town and country. Just as their respective central places are ranked, so a hierarchy of complementary regions is established. The author recognizes the difficulty in determining such "complementary regions" because they vary with the different types of goods and are subject to seasonal and periodic variations; finally there may be a certain degree of overlapping at their peripheries.

Distance is a major factor in the development of complementary regions, in particular the distance measured in time and cost, called economic distance. The range of a good is by definition, the maximum distance that a scattered population is willing to cover in order to procure a good supplied by a center. If there are two competing central places then this range is as small as possible. Moreover the development of central places depends on the population's net income. Thus the size of a central place is a function of three independent variables: the size of the complementary region, the volume of its population and its net income.

The central goods supplied by a larger central place have a greater range than those supplied by smaller places. As a wider variety of goods is supplied by a higher ranking central place than a lower ranking one it follows that the size of a central place has a direct influence on the range of a central good. Christaller remarks that giving buyers the opportunity to purchase, on one trip, a variety of goods from centers supplying them simultaneously, has the same effect as a general reduction in the price of goods supplied by larger towns. Hence each type of good has a specific range which differs according to the centers and the direction from a same center. That range depends on four groups of factors: the size of the center and the spatial distribution of the population, the price that buyers are prepared to pay, the subjective economic distance and finally the price and quantity of the good at the central place. Determined in this way, the range of a good around a center describes a ring. This ring has an upper boundary beyond which the good under consideration cannot be obtained from a given center. It also has a lower boundary inside which the minimum consumption necessary to cover the costs of production is not reached.

Generally, these upper and lower boundaries of the range of central goods have resulted in a system of central places comprising several types of size. According to Christaller this system is determined by three organizing principles each governed by its own laws. First of all the "market principle" states that in the case of a plain which is equally accessible from all directions, the complementary regions become hexagonals and those of the larger centers encompass the complementary regions of the smaller centers. The law which relates one or several higher ranked centers to the lower ranked centers it (they) dominate(s) is a geometric progression with as base one and a ratio of three, the base used being the highest ranking center. The relationships between size, spacing and commercial functions in this system give rise to hierarchy with seven ranks, going from market hamlets to regional capital cities, according to the data collected for South Germany. The strong correlation existing, on the one hand, between the size of centers and the frequency of central places and on the other hand between the size of centers and the volume of traffic should not detract from the true relationship of cause and effect. Christaller believes that size determines the other factors. In the system governed by the "market principle" it is assumed that all areas can be supplied by a minimum number of central places. A second principle called "traffic principle" assumes that the distribution of central places is optimal if the greatest possible number of large places is located along the routes connecting the largest centers, the cost of those routes being minimised. Then the complementary regions of the larger centers incorporate those of smaller centers. In this case the law relating the ranked centers to each other is a geometric progression with as base one and a ratio of four, taking as base the highest ranking center.

Finally the third organizing principle, called "separation principle" is essentially administrative in nature. It is based on the idea that complementary regions are separated for the purposes of protection and distinction which naturally implies a well-defined administrative control. The law relating the ranked centers to each other depends on the country's administrative organization. More often than not it is a geometric progression with as base one (the highest ranking center) and a ratio of six or seven.

It is these three principles, the first two economic, the third political which determine the system of central places. Depending on the circumstances, any one of the principles may prevail: a prevalence which in turn depends on short and long-term processes.

Christaller's study of South Germany gives good empirical support to this model. In agricultural regions, the market principle dominates whereas the traffic principle tends to prevail when the direction of commodity flows is not linked to their marketing.

Christaller's work which he expounded several times during the following years [81][94][101][118], not only sparked off a wave of controversies but had also a great influence, particularly after his works became known in the United States with the publication, in 1941, of an article by Ullman [117]. In the short run it paves the way for Losch's theory of economic regions. In the long run it can be seen as a decisive stage in the development of modern theoretical and quantitative geography.

Section 3. FROM FETTER TO OHLIN

In its critique of pure and perfect competition, neo-classic price theory was bound to confront the spatial factor. Nevertheless, both in the learned contributions of Joan Robinson and Chamberlin and in the well-known controversies appearing in the "Economic Journal" between the wars, the problem was never placed in its proper framework. The theory of "imperfect competition" saw the spatial factor only as a cause of market imperfection, while the theory of "monopolistic competition" integrated space within the larger framework of product differentiation.

Using the same assumptions and methods, contemporary spatial theorists studied the relationships existing between the sizes of markets, the location of the sellers, and the laws of price formation. Their contributions are numerous and scattered.

Already in 1901, J.B.Clark had posed the question of how market areas are determined, but he did not give the problem a clear and precise solution [21]. The analysis of Launhardt [16] had been forgotten at this time.

Fetter took up this question anew in 1915 [37]. He defined the point of separation between two markets through the relationship between prices and the cost of transport. Then in 1924, in a well known article, he approaches the study of the size and shape of market areas [44]. He shows that the boundary between the market areas of two competing sellers, separated by a given distance, is the locus of points, for which the difference of transport cost from the location of the two sellers is just equal to the difference in their prices. On either side of this line the difference in freight and in prices is not equal. Given the postulates of his analysis, especially that of the identity of the rate of transportation per unit of distance between all the points on the surface considered, the boundary is a hyperbola. The relationship between the prices determines its location.

Schilling calls this locus of points for which two or more sellers are in spatial competition the "isostante" [47]. He also introduces the concept of the "economic front" (Wirtschaftsfront), a line along which all prices are equal. The isostante becomes, in the case of competition between a rectilinear economic front and a point of production, a parabola if the transport costs are equal, and an ellipse if the transport costs are higher for the point of production.

A series of articles, which appeared mainly in the Journal "Technik und Wirtschaft", investigated the same problem. These articles introduce no new elements, except that the proportional rate of transport is replaced by the actual rates established by the German railroads and some new applications are attempted [42] [48] [85] [88]. Fetter also investigates the competitive relations between several metallurgical centers [70]. Isostantes are also used for a study of "hinterlands" [82].

Jonasson introduces the concepts of "isodistant", the locus of points at equal distance from a center; and of "isovector", the locus of points for which the cost of transport from a center is equal for a given product. By introducing the isovector in the tradition of the isodapanes of Weber and the isostantes of Schilling, he added supplementary elements to the isoline technique, which was later to be generalized by the Swedish spatial school (3).

However, these analyses require the assumption that prices are known and determinant in their influence on the size of the market, given the conditions of transport.

A second line of research, which was combined with Palander's work, was developed by Hotelling and embraced the relationships between the formation of prices, market areas, and locations of the sellers [65][79].

Hotelling begins with a criticism of Edgeworth's theory of competition among a few sellers. Contrary to Edgeworth's results, if one of the sellers continually raises his price while the others keep theirs constant, the buyer would reduce his demand gradually and not precipitantly. A revenue schedule appears which is the result of the discontinuous variation of the number of contracting parties with respect to the demand. Consequently, there arises a system of uneconomic and socially non-optimal prices and an exaggerated tendency of entrepreneurs to imitate their competitors. Hotelling is influenced here by the well-known article by Sraffa which showed the existence of subregions at the heart of the market within which quasi-monopolies arise [56].

He insists, thus, on a rigidity of demand which allows for the coexistence of several prices and the maintenance of a stable equilibrium. The cause of this rigidity which he considers most important is the cost of transport per unit of distance on a linear market along which the buyers, who have an inelastic demand, are uniformly distributed. Two sellers, behaving symmetrically, share a market at the point of indifference at which their prices and the cost of transport of their products are equalized. From this indifference equation Hotelling derives profit equations whose derivatives give the conditions for maximization. He obtains, in this manner, the optimal price and output of each seller. Furthermore, since his equations contain the distance factor, he can determine the optimal respective positions of the sellers by considering their locations as dependent variables. He then demonstrates that locational equilibrium implies the concentration of sellers at one point.

Chamberlin was to arrive at this last conclusion in the case of competition between two sellers; but, in contrast to Hotelling's generalizations, he shows how a dispersed location pattern, which is closer to a social optimum, replaces the clustered location pattern, when n sellers are assumed [75]. In the case of three sellers, two will locate at the quartiles and the third somewhere in the middle. For n sellers, if the length of the market is considered as unity, the space between sellers will be equal to 1/n (if n is even) or to 1/n+1 (if n is odd). This space will never be greater than 2/n when sellers cluster in pairs.

However, the less abstract nature of his original assumptions lead Chamberlin to extend his investigation to those cases where competitors have an advantage in being near each other. Thus he introduces an uneven population distribution and makes the length of the market a variable in the analysis. The distribution of the sellers appears then as a function of the distribution of population and the behavior of the buyers (shopping practices). The location clusters are in turn scattered according to the location factors of the particular industry.

The application to retail business of the theory of monopolistic competition causes the appearance of an urban rent different from agricultural rent (4). While the latter is a competitive revenue which varies as a function of the distance from the market, the former is a monopolistic revenue which depends on the size and nature of the market.

The whole set of analyses of limited competition led Austin Robinson, who was also influenced by Marshall's economic biology, to reconsider the importance of the size of firms, an idea which had been neglected since Weber's analysis of agglomeration [84].

Since the size of the firm's market depends on the density of demand and on the area which can be served for a given transport cost, the differentiation of the product within these boundaries depends on these costs; and the size of the firm is derived by its marginal receipts. In place of Florence's assumption of a limitless size, A.Robinson, along with Jones, substituted the concept of optimal size, because increasing cost implies the scarcity of one factor of production at least and entrepreneurship is scarce and unique by definition.

However, in 1931, E.A.G.Robinson had already given a more precise and comprehensive definition of the optimal size of the firm [72] (5). The size of the market limits, in effect, that of the firm; and the optimal-size firm is the one which has the lowest average unit cost per unit in the long run. The author distinguishes five determinant forces, from which are derived five optima: the technical, organizational, financial, commercial and flexibility optima, which he reconciles through joint control. From there he moves on to the concept of the optimal industry. The cost of production depends not only on the size, but also on the number of firms and the limits on efficiency due to bottlenecks in the supply of primary materials and to the increased cost of transport incurred in the extension of markets. In E.A.G.Robinson's presentation, the market itself depends on the location of national and international production. The analysis of the former restates Weberian theory in more modern terms and that of the latter, the classical nonspatial theory of the international division of labor.

It is at this time, moreover, that the sharp dichotomy in political economic theory by the evolution of price theory appears between the classical theory of international trade and the theory of price interdependence. Ohlin attempted to show, by introducing the spatial factor, that the theory of international trade was a part of a more general theory, the link between the two being the concept of region.

In 1924, he analyzed the formation of prices between two markets, called "regions", within which factors and products were mobile between which factors were fixed and goods mobile. The costs of transport and customs were not included, but their influence, added to the immobility of labor and capital, was reintroduced later as a complicating element [45]. Trade is generated from the differences in the regional endowment of the factors of production and the growing productivity due to economies of theory of international trade. At the end of his book, however, he suggests that the mobility of factors and goods is so limited that it is certain to influence the formation of prices. Therefore, the size of the regions should be reduced so that only local differences would be studied. The theory of interlocal trade would be a generalization of that of interregional trade, encompassing a large number of markets and maintaining not only the total sum of factors, but also their local distribution and their influence on prices.

The difficulties of this complex methodology lead to uncertainties and inadequacies in Ohlin's fundamental work of 1933 [77]. In the simplified theory of interregional trade, the region is deprived of all spatial significance by the maintenance of the classical assumptions concerning mobility and the costs of relocation. Ohlin's theory of "Buying Power" is a real advance when compared to the analysis of Mill and his successors. But it is not a contribution to spatial theory. The simplified theory of international trade, which assumes regions to be like different countries and completes the simplified theory of interregional trade by allowing for modifications in the supply of factors, considers only punctiform markets.

On the other hand, Ohlin introduces obstacles to the mobility of products and factors through the concept of "transfer costs", thus opening interregional trade theory to locational analysis within a framework of a "district whose boundaries are not described". A later article, reflecting the influence of Thunen, relates rent theory to that of the interdependence of prices [86]. Ohlin applies a modified version of the Isolated State to industrial production in the heart of his unbounded district. With a natural resource located at the center and a surrounding zone of agricultural cultivation, the land rent and prices of the products can be determined by a system of price interdependence. To this total approach, he adds a step by step analysis in Weberian style. Thus the theory of interregional trade can be thought of as a theory of location. The prices of the factors, the demand for goods at each point, and the costs of transfer being known, Ohlin deduces which products each region could provide most cheaply, for which other regions it will provide products, and in what quantities.

By introducing a monetary exchange system and international boundaries, a theory of international trade can be reached. The rate of exchange introduces a new factor and the need for having an equilibrium in the balance of payments requires a new equation.

However, because the theory of interregional and international trade is not based on a theory of location which does not apply to well defined districts, the connection between the two theories is not clear. The introduction of the concept of "subregions", intermediate in size between the region and the district appears to be of little use. This method treats spatial problems only as partial modifications of the results of interregional theory. Ohlin's analysis therefore fails to provide the link between the theory of location and the theory of exchange.

It was not until Losch that the theory of trade (Handel) was for the first time integrated into spatial analysis. Palander's model was developed only for a closed economy. Nevertheless, by shedding some light on the close relationships existing between all the economic variables and by initiating the integration of location theory into the framework of modern price theory, Ohlin's work was among the major influences on Palander.

(1) Bortkiewicz (L.) in Deutsche Literaturzeitung, 31, 1910, 1717-1724, Schumpeter (J.) Jahrbuch fur Gesetzgebung, Verwaltung und Volkswirtschaft, 34, 1910, 444-447.
(2) Englander insists that rent is a function of location and not a "direct" factor of location.
(3) Jonasson (O.), Jordburkets beroende avdet geografika marknadslaget Kungliga Lantbrucks akademiens Handlingar, 1930.
(4) Chamberlin (E.H.) [75] Appendix D.
(5) Hypotheses of a more dynamic character had already been expressed by Epstein (R.C.) since 1929. Cf. Epstein (R.C.) "Locality Distribution of Industries", The American Economic Review, 19, Suppl. March 1929, p.172ff.

Chapter 6. Tord Palander

INTRODUCTION. THE GOALS AND METHODS OF PALANDER'S THEORIES

The Swedish economist, Palander, made the first major non-German contribution to spatial economics in his 1935 thesis [87].

In conformity with the contemporary approach of both orthodox theoreticians and specialists in spatial economics, Palander tries to develop a theory of general equilibrium.

He starts by exploring, in general terms, the problem of the spatial distribution of production and markets in capitalistic countries. In an economy with division of labor, the spatial problem arises from the possibility of locating production at some point other than the place of consumption. While in a family economy, Palander claims, men adapt themselves to the geographical conditions.

This realization leads him to classify the spatial relations of economic life in a particular way. The technical relations between production and consumption include both production linked to the place of consumption (Konsumgebunden) and production independent of the place of consumption (Konsumlosisch). Technological evolution concentrates production and decentralizes services, which are nearer to the consumers. Since the final product is spatially tied to the consumer (Konsumgebundenheit des Fertigprodukts), the problem of location, Palander concludes, concerns only the earlier stages of production. Similarly, the technological relations between outputs and factor inputs lead alternatively to production that is spatially tied to those places where one or more of the factors is obtained (lagergebunden) and to production that is spatially independent of the factors sources (lagerlosisch). If production is tied both to the location of consumption and to that of the factors, as in the case of tourism, it is the consumer who will move. Generally, because of technological progress which allows the substitution of more easily transportable factors for naturally fixed factors, the location of production has become increasingly independent of the location of these factors. Since it is difficult, in any case, to determine whether final production is completely tied to its factor sources, Palander proposes that location should be designated as factor-oriented only if the movement of intermediate product inputs is absolutely impossible. Thus the proportion of production that is input-oriented depends on the facility with which a transportable intermediate product input can be obtained and on the number of stages of production for which the use of a given fixed factor is necessary.

He distinguished, moreover, between those factors that are capable of only one use and those with multiple uses. For the latter, their use may be alternative or simultaneous. The technology of production, in the meantime, can be fixed or can allow for possibilities of substitution. If no substitution is possible, the location of production will depend on the spatial distribution of the input sources. In the case of substitution, the same product can be produced with different combinations of factors depending on the relative prices of the factors at different locations. Thus, the theory of location must take into account the methods of production.

Palander concludes that a whole series of problems of location exist. Several categories of economic calculations appear according to the nature of production. For activities linked to consumption, the question is which factors to use; for activities linked to natural factors suitable for only one use, the question becomes which factors and what method of production to use. For activities linked to natural factors that have several uses, the question becomes that of the choice of the factors, products, and method of production. For those activities that are independent of the factor sources, it is the choice of the products, place and method of production that is in question.

Palander introduces considerations on mobility of the factors. Some are transportable in the same manner as the finished products; e.g. capital, power, raw materials, machinery. Others, although they can be moved, may only do so slowly; e.g. fixed capital or labor. The location of the latter is related to housing conditions and commuting patterns (Pendelwanderungen) of workers over a surface whose dimensions depend on the means of transportation and the time the workers are willing to spend away from their homes. These movements may be daily, weekly, or seasonal. But, Palander believes that labor is always a fixed factor in the sense that it only moves for extra-economic reasons.

In a general way, Palander recognizes the fact that climatic, governmental, and institutional elements influence location.

By this combined examination of the questions posed by spatial differentiation, Palander shows that the location problem cannot be solved by a single computation, but rather that it implies several series of alternative calculations. He thus provides the first generalization of the problem which later was to be systematically analyzed by Losch.

The location of consumption raises complex problems because it is not always determined by the location of production. At a time when the capitalist was the entrepreneur, he was tied to the enterprise, which was itself tied to labor. Financial and technical evolution made capital and entrepreneurship independent of each other leading to the independence of the capitalist and the enterprise. Stockholders, pension holders, landlords, and those who receive rent from natural resources are also independent of production. Furthermore, the location of administrative personnel, according to Palander, has no relationship to the distribution of economic activity. Similarly, large enterprises have an administrative sector that is independent of the place of production. The transportation, buying, publicity, sales, and other administrative departments of large enterprises are influenced by a combination of economic motives to locate in cities.

In this manner Palander sets out the framework of a general theory of spatial interdependence of units in the economic system. He felt that he was extending the work of earlier location theorists. Due to the formal character of his work (a thesis) a long descriptive and critical history of spatial economic theory was presented. Palander had the opportunity to set up a balance sheet comparing previous contributions, while at the same time presenting these theories to the Swedish economists. His historical account, which is not complete, sacrifices the search for general relationships in order to provide a precise and pedagogic argument. This argument is restricted because of its lack of critical discussion or its preoccupation with details. Palander distinguishes between "special theories" and "universal theories" of location. Some of the special theories are related to the analysis of

land rent, which involves the criticism of the works of Ricardo and Stuart Mill, and leads to a study of the work of Thunen. Other special theories concern the analysis of the location of industry, found in the work of Adam Smith, Roscher, Schaffle, Ross, Hall, Launhardt, and ending with Weber; still others concern the analysis of the size of the market, a subject which Launhardt, Fetter, Schilling, Schneider, etc. discussed. Finally, the spatial price analyses of Launhardt, Hotelling, Schneider, etc. are covered. It is here that Palander, who was generally in contact with these contemporary works, surpasses the achievements of previous authors and introduces most of the fundamental equations of his own theory. Coming back to the "universal theories" of Predohl and Ohlin, Palander establishes as his objective the determination of guidelines for a synthesis, particularly by retaining only the local differences existing within each system and abstracting from
external relationships. This discontinuity in the exposition is explained by the fact that the author, in his search for a general theory, devotes his efforts mainly to the last type of the "special theories" mentionned above, which he then tends to generalize.

Having posed the general problem of location, Palander, in effect, ties it to the theory of prices. But the models of Walras, Cassel and Pareto, valid for a punctiform market, are not able to explain the location by reintroducing factors dependent on local variations. In effect, Palander raises three fundamental questions. First, the assumption of free competition presupposes the existence of a single market with an unlimited number of competitors and the sale of only one product. But, the absence of a single market, together with local variations in prices, are a continuing and essential characteristic of spatial price formation. Second, the simple relationships between cost and price, as well as the marginal relationships between factor prices and the value of products, are no longer valid when there are permanent local differences. Finally, the theory of prices deals with static equilibrium and does not explain the dynamic variations of location.

The example of imperfect mobility of labor or real capital is sufficient to challenge the idea that the factors of production are instantly attracted to the places where their returns are highest. The spatial distribution of economic activities depends, on the one hand, on the conditions prevailing in previous periods, and, on the other hand, on the speed of movement of factors of location. It is therefore important to place the study of location in a time dimension, so as to take into account lags in adaptation. The analysis must incorporate the reaction time of the diverse factors of location.

A general explanation of the location of production and factors must be dynamic, starting from a given state and the other elements being technological and institutional conditions, schedules of demand for products and of supply of factors, dynamic evolution of demand and supply, and the speed of movement of factors. Since the latter elements are unknown, such an analysis becomes impossible; moreover there is no dynamic theory whose validity is generally accepted.

Therefore Palander limits his scope and proceeds to study the reactions of the entrepreneur in the face of local differences in economic conditions and then to relate the results to what is known about the movement of factors and the variation of technology and demand so as to obtain a general explanation of the process of evolution. The results of this process with regard to location depend, in effect, on the following factors: the situation at the beginning of the evolution; the adaptation of the firms through time; the movement of the factors of production; the movement of the centers of consumption; the variations in technology and the variations in institutions. In the short run, the last four elements are constant; and, the shorter the time period, the more important is the first factor.

Palander, however, limits himself to an analysis of the second factor - adaptation over time.

Entrepreneurs adapt to changing conditions in five different ways: first, in the creation of new enterprises; second, in the creation of branches or the relocation of old enterprises; third, in the adaptation of old firms to changes that are due to modifications in investments, prices, sales methods, etc.; fourth, in the closing down of some old enterprises; and fifth, in the change in the relationships between enterprises: mergers, partitions, acquisitions.

Previous theories neglected the internal changes which occur in firms, despite the importance of such changes. The theory of location must start from a general theory of production and be extended to cover local differences.

It can be understood, then, how Palander was led to study the reactions of the entrepreneur to market conditions as he sees them, and to analyze the way in which certain factors, such as transportation, local cost differences, and local markets, jointly determine locations.

He announced that his investigation would be conducted in five stages: a study of the cost of transport under certain simple conditions; a mathematical study of the location of an enterprise and of the choice of production processes with respect to the location; a study of the influence of local methods of price determination and the interpretation that the entrepreneur makes of the price policies of competitors; the location pattern of production, and the location of new enterprises; a study of the various adaptations made by the entrepreneur during the course of dynamic evolution; finally, an application of the results obtained to the case of agriculture.

But it is here that another limitation on the scope of Palander's theory arises. Palander's work is incomplete; it comprises only the first and third of these studies. The author was satisfied to announce the publication of the other three in the "Economisk Tidskrift". Thus, discontinuities in the theoretical reasoning itself are added to the fact that the beginnings of the analysis are introduced in the historical part of his work. This means that Palander combines a broader conception of the problem of location with a series of peculiar analyses that are linked, at best, only by the unity of the mathematical framework.

Section 1. MARKET AND LOCATION

Palander notes that the theories based on Thunen and Weber assume that the buyers' and sellers' markets are independent of location. But, the spatial distribution of buying and selling among various parties can be studied from two points of view: one, the dependent variable will be the size of the market, given the price, location, cost and possibilities of transportation; or two, the analysis will comprise the relationships between price formation and the size of the market.

In the first case, Palander follows Launhardt and Fetter. He borrows from Launhardt the assumption of two locations separated by a given distance and competing for the surrounding market. The buyers pay the factory price plus the cost of transport, which is proportional to weight and distance. By assumption, buyers choose the product of the producer who provides the lowest delivered price (1).

Thus there are points for which the sum of the prices and transport costs for the products of the two sellers are the same. The equation of the locus of these points gives a fourth degree curve which is generally closed and which marks the separation of the two markets. Palander, following Schilling, calls this an "isostante". Inverted cones are placed on top of a graph representing the surface of the markets, their vertical axes passing through the location of the sellers. The slope of the sides of the cones expresses, geometrically, the local prices as they increase with distance. The intersection of the surfaces of the cones constitutes the locus of points for which local prices are the same for any two sellers. The isostante line is simply the vertical projection of this intersection.

In certain simplified cases, the equation of the isostante is simpler than a fourth degree equation. Palander mathematically demonstrates that if the prices and transport rates are the same for both sellers, the two markets are separated by a perpendicular line bisecting the line that joins the two locations. If only the prices are the same, the isostante is a circle. If only the transport rates are the same, the isostante is a hyperbola. If the two sellers are located at the same point, the isostante is again a circle and the more expensive product could only be sold at a distance from the place of production. Finally, if the market is linear, that is reduced to the line that joins the two locations, the point of separation of the two markets is where the algebraic sum of the prices and transport costs are the same.

By his various mathematical models, Palander proves that in the third, fourth, and fifth cases, the product that has the lower transportation rate cannot be excluded from the market whatever its price. He also shows how a proportional reduction in the transport rate enlarges the size of the market of the less expensive product, while a proportional rise in the transport rate enlarges the extent of the market of the more expensive product.

When the prices are no longer given, the analysis becomes complicated. On this fundamental point, Palander follows Launhardt and Hotelling. Once again he adopts the basic assumptions used by his predecessors (2).

Consumers, whose demand is inelastic, are uniformly distributed along a line of given length, and on which are located two competing producers. Buyers pay a delivered price and buy as cheaply as possible. The cost of transport is always proportional to the distance.

There is a point for which the sums of the unit costs of production, profit and cost of transportation for the two sellers are equal. Given these assumptions, the quantities sold by each seller are a linear function of the distance to each consumer. Total profits are defined as the product of the quantities sold times the unit profit. Palander gives this a mathematical formulation which includes the distance factor. In his equations, the locations of sellers are given and their hinterlands are also given, while the lengths of their contiguous markets are variables.

The equations of profits show that the profit of each seller depends on the profit of the other, since each equation contains the unit profit of both competitors. Thus Palander develops a duopoly theory of double dependence. Each seller fixes his price assuming that the price of the other is independent of his own: i.e., that his unit profit is constant. Total profits are maximized by setting the partial derivatives of total profits with respect to unit profits equal to zero. The equations can then be solved to find the equilibrium prices.

Palander's formulation includes the models of Launhardt and Hotelling as special cases, since Launhardt located the sellers at the extremes of the market, thus eliminating all hinterland, and Hotelling imagined a zero cost of production and equal transport costs.

The author easily determines the price of each competitor by supposing the transport cost equal and the cost of production nil. This permits him to identify, in the preceding equations, the prices as equal to unit profits. A given profit can thus result from a series of combinations of different prices. In a coordinate system where the prices are indicated along the axes, Palander constructs a graph of the curves indicating all the price combinations corresponding to given profits; the equations indicate that these constant profit curves are hyperbolas.

Any variation in prices is expressed on this graph by a horizontal or vertical shift. Each seller tends towards the highest profit possible, given the price of his competitor. At each competitor's price, there is an optimal unit profit determined by the tangency of the competitor's price and the seller's profit curve. The locus of these tangency points gives the line of adjustment for each seller. At the point of intersection of these two "lines of adjustment", a stable price equilibrium will be reached.

However, these "lines of adjustment" have economic meaning only within the limits imposed by certain restrictions. In effect, if one of the sellers is able to fix a price lower than that of the other by at least the cost of transportation between their two locations, he will eliminate his competitor from the market. From this conclusion Palander obtains the equations of two new lines which determine the limits of the "feasible region" of competition (Mogliches Feld).

Thus the author is in a position to make more precise an analysis of market strategies. According to whether one of the sellers adjusts his price following his "line of adjustment" or whether he adjusts it so as to dominate the whole market, Palander obtains two different total profit equations. It can be seen, by inspecting these equations, that following the second policy will be more profitable if the price of the competitor is high, while the first policy will be more profitable if the price of the competitor is low. The final choice will thus depend on the particular value of the other seller's price and this can be determined mathematically.

The possibility of such a strategy can lead to a perpetual disequilibrium of price. A stable equilibrium can only be reached if the two lines of adjustment cut each other at a point where they effectively determine the prices for both sellers. But, given the price equations in the case of equilibrium on the one hand and in the case of the elimination of a competitor on the other hand, Palander determines certain lengths of "hinterland" beyond which the "lines of adjustments" would no longer guide the policies of the sellers. In other words, when a firm has an extended hinterland, the competing firm, instead of sharing the intermediate market area, prefers to lower its price and capture the whole market, including the hinterland of the other firm. The equilibrium finally depends on the location of the sellers and the length of the linear market.

Section 2. TRANSPORT COST, LOCATION OF PRODUCTION, AND MARKET SIZE

In his approach to the analysis of location, Palander suggests that location factors be analyzed one by one, while holding all others constant. The analysis should start with the cost of transportation on which the others depend.

To make the constants and the variables of his model more precise, Palander shifts his analysis to the level of the firm and to the level of the market; he distinguishes two problems: either the prices, the locations of raw materials and energy, and the places of consumption are given, and the location of production is sought; or else the production point is given together with the conditions of competition (locations of enterprises, factory prices, transport cost paid by the buyers of finished products or the sellers of raw materials) and the influence of the price on the size of the market is studied.

From the three elements that constitute transportation cost (loss of time, loss in the value of goods, and shipment costs), Palander considers only the latter. In order to measure the effects of transport cost, he defines more precisely the factors which influence these costs.

In the case of regular shipments where the prices are generally fixed by public control, the various price policies lead to substantial variation in costs.

The costs of transport are tied, first of all, to the type of transportation. Costs can be independent of distance, as in the case of postal rates or when the rate is fixed in accordance with the volume of traffic. They can be proportional to distance, following a uniform and (or) variable rate; or finally, they may vary with direction. Transport rates also depend on the nature of the goods shipped. Generally the costs are proportional to weight, but the specific value of the transported product can also be taken into account. In this case, Palander suggests returning to the Weberian concept of "ideal weight" which could be used in all those cases where the rate is the same for raw materials and final product. Rate variation according to the value of the product modifies some of the theoretical results obtained previously. On the one hand, even if the raw materials weigh more than the final product (that is, in Weberian terms, if the "material index" is high), higher transport rates on final product due to higher value can throw off the relationship between the "ideal weights". On the other hand, it is no longer important to know which of the raw materials loses some of its weight during the process of production; only the sum of the "ideal weights" of the raw materials is important. Transportation rates can also be related to the volume of shipment; costs are lower for full carload lots. The existence of contracts and personal contacts between producers and transport firms can complicate further the analysis by modifying transport rates.

In the case of irregular shipments, the price is fixed by supply and demand. The relationship between price and distance becomes uncertain. Often the price is determined not as a function of the one way distance, but rather of the round trip, including a consideration of the cost of return trip, the season of the year, the extent of the circuit, and the capacity of the means of transport.

With a view towards the integration of these diverse conditions of transportation into his abstract analysis, Palander generalizes and synthesizes the descriptive method of isolines, a technique which consists of joining, with lines, points possessing the same characteristics. Palander borrows from Weber the notion of isodapanes, lines of equal total transportation cost. From Jonasson, he borrows the idea of isodistants and isovectors, the latter being lines of equal transport cost for a given product. From Schilling he borrows the idea of isostantes, which unite all the points where two or more enterprises located in different places can offer the same delivered price. Isotims, finally, join the points where the delivered prices of a given product from a specific source are equal; and isochrones unite points of equal shipping time (3).

The process of abstraction includes, in addition, the definitions for the remaining concepts, not yet made precise. Palander calls a region where all points are interconnected by a given means of transportation a "transport surface". A "transport line" connects a set of points; and a "transport point" identifies a characteristic place (a terminal, port, point of trans-shipment, etc.).

A transport surface is formed in reality by the sea or by a region extremely dense in transport lines. If the cost of transport depends on the distance and if it does not involve any discrimination or discontinuous variations, the isovectors, assuming a uniform rate, are concentric and regularly spaced. Or, if a variable rate is assumed, they are still concentric but their separation becomes larger the farther away they are from the center. Representing as they do the cost of transportation from one place to another, they are of special interest with respect to location if the market is analogous to that of Thunen's "Isolated State". Then they become the same as the isotims for the products whose price is fixed at the central market. If more than one shipment is needed, as is the case with most production, the isodapanes are most useful. When there are two shipments on a transport surface, the isodapanes will go through all the points of intersection of the isovectors for which the sum of transport costs is the same. For three shipments, isodapanes for two of the shipments are constructed, then the final isodapanes will unite the isovectors of the third shipment with the isodapanes of the first two; and so on for more than three shipments.

The series of isodapanes drawn by Palander illustrate certain hypotheses. Thus in the case of two shipments, if the raw materials weigh twice as much as the finished product, the graph shows that the point of minimum transport is found at the place where the raw material is extracted, if the transport rate is uniform. Cost of transport of the raw material rises with the distance from the extraction point. On the other hand, from the consumption point, the cost of transport of the final product rises at a slower rate. If the transport rate is variable, the cost rises rapidly in all directions, but at a certain distance from the place of extraction of the raw material and of the place of consumption there is a zone where the cost becomes more or less constant. The variable rate thus tends to create several minimum points of transport cost between which there are zones where the cost shows only weak variations. In this case, the way of determining the optimal transport point employed by Launhardt and Weber is no longer valid. The isoline technique allows Palander to solve a problem which Weber had already faced - the problem of minimizing the rise in transport costs caused by the deviation of production from the minimum point due to the change in relative costs of factors of production. Depending on the nature of the rate and the relationship between the various weights and points of production and extraction of raw materials, location can vary within certain limits and even, in some cases, throughout the whole figure, without having more than a 20% increase in the cost of transportation. Therefore within certain zones the transport factor ceases to be the essential factor of location.

This possibility of displacing the point of production within certain limits often involves a choice among several sources of raw materials. Palander constructs isodapanes so as to unite the points where the costs of raw materials plus total transport costs are equal; the ultimate isodapanes are obtained by putting together several simple systems of isodapanes.

However, in the case of the transport surface, two difficulties can modify the isolines without modifying the method of reasoning. Irregularities in the nature of the transport surface (detours, impossibility of transportation at certain points, etc.) distort the isovectors, the isodapanes and the whole model of location. The convergence of several lines at the same point (isthmus, channel, bridge, etc.) gives birth to "agglomerations caused by the conditions of transport" of a certain number of independent locations. Even on a continuous transport surface, discontinuities due to the costs of loading, unloading, etc. can be found. The tendency for production to locate at the extremes of the 'locational figure" is reinforced. Similarly, a separation can cut the transport surface in two (difference in the gauge of the railroads, political boundaries, etc.) and can reinforce the attraction of location along or near this boundary.

In the case of transport lines on which the cost of transport is clearly lower than in the rest of the surface (canals, sparse railroad networks, etc.), points of transport intersection appear to be the most advantageous locations.

In the case of transportation between two locations situated on two different transport surfaces whose rates are independent, Palander proves mathematically that if the costs are the same, transportation takes place in a straight line and if the costs are unequal, transportation on the more expensive surface approaches the perpendicular to the line of separation of the two surfaces more closely as the cost differential increases (4). Thus in the case of a coast line, the product is sent through the port that increases the overland distance the least. If the modes of transport are combined on both a line and a surface (the case, for example, of a canal crossing a region where there is a high density of roads), Palander reduces the problem to that of the preceding case because the system will behave as if one of the two surfaces were reduced to the line separating the two surfaces. Clearly, if the straight line which unites the two locations approaches the perpendicular to the boundary line, the position of the locations is not indifferent to the solution of the problem. Direct transportation is then more advantageous than combined transport.

Finally, Palander converts the cost of trans-shipment into an increase in the length of the distance to be covered; and he constructs lines containing points for which the costs by direct and combined transportation are equal. There isolines break the transport surface into zones where the combined transport is more expensive or more economical with respect to a given point.

However, no matter what the complications of the analysis, they can be reduced to distorsions in the form of isovectors and modifications in the results without affecting the approach and the method of reasoning.

Finally, Palander applies the results of his study to the case of the combination of transportation by ship and rail and to the competition between rail and highway transportation.

Land and sea are thought of as two transport surfaces with different costs, the land being traversed by cheap lines of transport (waterways) butting across a surface of transportation (a dense network of railroads). The saving in cost of water transportation is so large that the theoretical distance between points located on the coast or on a canal becomes very small. The separation between land and sea is not a straight line, but a series of points, seaports, whose location is determined by the shape of the coast and is generally in bays or estuaries. The port determines the network of transportation of which it is the center. It becomes the point of minimum transport cost for numerous industries and its advantages with regard to location have a cumulative character.

Competition between rail and highway transportation depends on a number of possible conditions. While railroads are characterized by monopolistic pricing, where fixed costs are important and rate discrimination prevails, highway transportation is characterized by competition and variable cost and rate largely proportional to the weight (not value) of the product shipped and to distance (uniform rate). Thus competition arises where distances are short and values for products relatively high. The flexibility of highway transportation limits the generality of any theoretical study, since there is an absence of tariff regulations as well as a possibility of taking freight on the return trip. There are other advantages that might also be present, such as home delivery, no trans-shipment, greater speed, etc. Palander assumes that road transportation takes place on a transportation surface cut by other lines of transportation (railroads). In addition, in most cases, the lines of transportation are reduced to a series of points (terminal stations). Using isolines, Palander obtains the places for which a simple or a combined means of transportation will be preferred. In this way, he derives important conclusions. If the transport line is reduced to a series of points, rail transport is the less advantageous means and transportation between two successive points is the most expensive. If the series of points can be approximated by a line (the stations are very near one another), the surface on which highway transport is preferred is larger if the point from which the transport is made is not located on the line. The surface on which railway transportation or combined transportation is preferred is larger if there is no trans-shipment. Finally, if the difference between the rail and road transport cost diminishes, road transport is preferred at a greater number of points.

The recent development of highway transportation did not fail to have an impact on location. In agriculture it diminishes the local differences in cost and brings about an increase in the activity of regions formerly too removed from the market via other means of transportation. For industry, the effect is to weaken the former tendencies for new firms to locate at raw material, market, or transport intersection points. In general, highway transportation encourages decentralization for several reasons: the existence of uniform transport rates; the combination of modes of transportation, which permits locations off of the major lines of transport; and the equalization of cost for final products and raw materials, which diminishes the attraction of the place of consumption.

Modifying his scale of observation, Palander goes on to show, in a brief analysis, that once the location is known, the price influences the extent of the market. Here, the location, instead of being a variable, is given, and the size of the market is the dependent variable. A particular category of isolines, the isostantes, is appropriate for this type of research. While Palander had, in previous cases, shown the mathematical determination of isostantes, he proceeds in this case to a geographical analysis. He shows how an analysis under the assumption of a single price is insufficient, and demonstrates the usefulness of investigating a whole series of interconnections between the prices of two or more firms.

When the prices of different competitors are equal, the isostantes become the same as the isodapanes. If the prices differ, the isostantes become the locus of the points for which the difference in transport cost is equal to the price differences. While the isodapanes link the points for which the sum of the transport costs are the same, the isostantes link the points for which the difference in total cost is the same. By means of a series of graphs, Palander shows how the isostantes allow the limits of two or more competing markets to be determined in the cases of variable or uniform transport rates. His method of approach is limited only by the existence of inelasticity of consumer demand, product differentiation, etc.

Section 3. CONDITIONS OF COMPETITION AND LOCATION

The last portion of Palander's work is concerned with the relationship between markets and locations. His analysis here is devoted to: the effect of competitor's pricing policies and of the local methods of the price determination on the prices, the local distribution of production and the location of new enterprises. While in previous analyses, included in his history of spatial theory, he had assumed that two competitors follow a policy of dependency ("autonomous policy"), his research now turns to the case where first one of the competitors, and then both of them, take into account the reactions of the other ("super political" behavior). Everything takes place as if Palander had juxtaposed spatial theories of asymetric duopoly and of duopoly of double control to a spatial theory of the duopoly of double dependency.

Once again he starts from Launhardt's and Hotelling's assumptions of two competitors, a linear market, an inelastic and uniformly distributed demand, and a zero cost of production (5).

In the case where one of the sellers decides to behave "super politically", Palander, connecting his mathematical formulation to the previously developed model, shows that this seller departs from the adjustment line of his competitor who has an "autonomous" behavior. When a seller takes his competitor's reaction into consideration, he fixes a price which, after the change in the price of his competitor, corresponds to the point on the adjustment line that touches the highest profit curve.

The adjustment line cuts the profit curves in such a way that profit goes up when one moves outwards on the adjustment line up to a point where this profit is maximum. The profit is thus at a maximum when the competitor fixes his price at the point corresponding to the adjustment line which is tangent to a profit curve. The derivatives of the profit equation of the "superpolitical" seller and the derivatives of the equation of the line of adjustment of the "autonomous" seller give the equation of the locus of the points of the profit curves which have a tangent parallel to the line of adjustment of the competitor. At the intersection point with this new line, the adjustment line of the competitor is thus tangent to the highest profit curve possible and the coordinates of this point give the respective prices of the duopolists. These prices define a state of equilibrium. Palander also shows that these profits are higher than those that would be obtained if the two sellers were to practice an autonomous policy and that the difference in profit is larger for that enterprise which continues to practice an autonomous policy.

This equilibrium is stable only as long as the adjustment line determines the policy of the "autonomous" seller up to the point corresponding to the maximum profit of the "superpolitical" seller. At that point the "autonomous" seller may prefer to lower his price to gain control of the whole market, and the equilibrium is then broken. Palander thus reverts to the equation expressing the seller's price at which his competitor shifts from sharing the market to attack or defense. He then deduces the maximum size of the hinterland of the "superpolitical" seller which would still be compatible with a state of equilibrium. Beyond this boundary, this seller will either set the highest possible price corresponding to the adjustment line of his competitor, or he will set the highest price which will not cause his competitor to go from defense to attack. The choice of policy is made in consideration of the size of his hinterland which is determined by the equations expressing these two alternative possibilities.

No equilibrium is possible if the two sellers both behave superpolitically, each adapting his own price policy according to the other's adjustment line. Palander, in order to analyze this case, could only assume that the second seller's behavior is identical to that of the first. His profit curves are also tangent, parallel to the adjustment line of the first. The locus of these tangent points gives a new equation which expresses the set of his competitors' prices for which the second "superpolitical" seller maximizes his own profit.

Starting from any given price, the duopolists react simultaneously to each other's adjustment line. Neither practices an autonomous policy, but each believes that the other one does. Thus, the price oscillates continuously; it is in perpetual disequilibrium. Eventually, each seller realizes that he has made an incorrect assumption about his competitor's policy and revises his own policy. Only then can an equilibrium be reached. Nevertheless, Palander believes that his calculations would be more likely to be revised on the basis of less fundamental issues, such as the cost of production. Then each duopolist estimates the new adjustment lines of his competitor and price oscillations continue within these new limits. If each believes that the costs of the other are higher, for example, then prices will rise continually and it becomes unlikely that the choice made on the basis of the estimates will remain valid.

Varying slightly his model to account for the fact that the conditions of competition depend on the customary methods of fixing prices, Palander corrects his equations so as to integrate into his analysis the cases where delivered prices are not f.o.b. price plus transportation costs. The reasoning is once again conducted with the assumption of a duopoly of double dependency, but it can be applied to the cases of asymmetrical duopolies or of double control.

First, one of the sellers can calculate his delivered prices from some location other than the place of production. His price is then equal to the price at this "base point" (Frachtgrundlage) plus the cost of transportation from there to the consumer. The equation of his adjustment line, which contains the distance to be covered, is thus modified to express the deviation from the point of production to the basing point. The equilibrium price is therefore modified. The delivered price of the seller who continues to use f.o.b. pricing at the point of production stays the same, while the price of the other seller goes up. The larger the distance between the base point and the production point, the higher the new equilibrium price will be. The same is true of profits. Finally if the location of the base point is variable, it will tend to approach the point of production of the competing seller.

The underlying assumptions, made by Launhardt and Hotelling are incompatible with the assumption that both sellers charge f.o.b. price plus transport. This incompatibility is caused by the impossibility of establishing an equilibrium division of the market in such a case. However, Palander could still investigate the case where only one of the sellers incorporates delivery cost and charges a uniform delivered price throughout. Once again there will be a point of indifference separating the market areas at the place where the two prices, no matter how they are calculated, are equal for the buyers. Following his usual methodology, Palander deduces mathematically the adjustment lines of the duopolists, the equilibrium prices, and the profit equations. The profits are lower than those obtained when the prices are fixed at the point of production. Therefore knowledge of the new equilibrium point would never motivate a seller to practise f.o.b. plus transport pricing. Nevertheless, Palander shows, by giving numerical values to his variables, that a seller will enlarge his profits as long as his competitor's reactions have not taken place. The determination of a f.o.b. price plus transport is of immediate interest only.

Finally, the location of the sellers, formerly considered as a given variable of the analysis, is now considered the dependent variable. Palander links the choice of location to price and to the method of its determination as well as to the cost of production, the cost of transport, and to the reactions of the duopolists. If a seller who is trying to determine his best location practices an autonomous policy, his choice will depend on his market strategy. Starting from the equations representing the different possibilities (conquest, defense, or sharing), Palander is able to determine the extent of the hinterland of the seller. With the assumption of the conquest or the sharing of the market, the profits increase as the hinterland grows, so that the seller tends to move closer to his competitor. With the assumption of defense, his location depends on the price of the competitor. When this price reaches a critical level, the tendency towards concentration disappears.

CONCLUSION

Once the assumptions of perfect competition are discarded, the simple complementary principles which structure general equilibrium theory and lead to marginal cost pricing of products and marginal revenue product pricing of factors must be abandoned. It is logical that this would differentiate Palander's work from the Walrasian tradition, in favor of the tradition of Launhardt, Fetter, and Hotelling.

It is not until Losch, however, that a complete system of equations is determined to express the spatial relationships of a general economic equilibrium under limited competition. Palander does not proceed beyond a static and partial analysis even though he examines the complex spatial interrelations of a closed economic system and emphasizes the importance of dynamics.

---

(1) Cf.Mathematical Appendix III,1.
(2) Cf.Mathematical Appendix III,2.
(3) Cf.Mathematical Appendix III,3.
(4) Cf.Mathematical Appendix III,4.
(5) Cf.Mathematical Appendix III,5.

Chapter 7. From Palander to Losch

The impact of Palander's work in partial spatial equilibrium had not yet been fully felt, when Losch began to elaborate a general spatial equilibrium theory. Only five years separated the publication of Palander's findings from the first edition of Losch's monumental work. During these years, the Anglo-Saxons continued the exploration of the spatial structure of markets. At the same time, the evolution of the world situation led to some new and original lines of spatial research.

Lerner and Singer return to Hotelling's model and examine the conditions under which it is valid [97]. They relax some of the basic assumptions in order to reach more general conclusions. The assumption of perfect inelasticity of demand, especially, allows an infinite rise in prices, an event unlikely in the real world. Therefore, the existence of a maximum limitation on price rises appears to be important for the determination of equilibrium prices, locations and profits. But the analysis does not reach complete generality until a later exposition by Smithies.

Hoover takes the first steps in the analysis of monopolistic competition and price discrimination, leading directly to Losch's later price theory [96]. Hoover develops a model expressing the relationship between spatial demand and marginal revenue. In general economic theory, price exceeds marginal cost by an amount which is a function of the elasticity of the demand curve. The prices must then vary if the elasticities of the demanders vary, discrimination being practiced against the buyers whose elasticity is lower. In spatial economic theory, Hoover demonstrates mathematically that there is a tendency for f.o.b. price to increase when the unit cost of transport rises. Therefore, spatial price discrimination has most effect on buyers who are farthest away. However, the author remarks that in fact, price discrimination often takes place in an inverse manner, with the sellers practicing a policy of freight-absorption which penalizes the nearest buyers. This reversal may be due to a greater elasticity of demand of the farthest buyers, to a rapidly decreasing elasticity when the prices are lowered, to a large number of seller locations which make the boundaries of the individual market areas more vulnerable, or finally to the existence of agreements or common controls, such as market sharing agreements, price agreements, basing point systems, etc.

Singer discusses Hoover's analysis in Chamberlin's terms of "general tax" and "differential tax" [98]. By altering the price competition, the monopolistic power imposes on the buyers a general tax measured by the excess of the f.o.b. price over the marginal cost. He imposes a differential tax measured by the excess of the average f.o.b. price in an imperfect market over the f.o.b. price imposed on the nearest buyer. While the general tax is just a distorsion of the price system, the differential tax is a distorsion of the spatial pattern of prices.

Finally, Hoover synthesizes those theoretical contributions of his predecessors which appear to have practical value. Using empirical data obtained from the shoe and leather industry, he is able to explain an actual location pattern [95]. Dean, in a similar manner, makes extensive use of historical data and contributes another major link between theoretical analysis and practical studies [102].

The wide fluctuations in the business cycle, which posed so many new problems to economists, did not fail to influence the evolution of spatial economics. At the same time, these problems required a better adaptation of theory to reality; they emphasized the necessity for the development of a dynamic theory; and they increased the awareness of the need for a link between time and space in spatial economics.

In order to analyze the changes in the most important industrial districts of the United States from 1869 to 1929, McLaughlin compiled statistical information on employment, demography, wages, value added, etc. for 33 "industrial areas" of the U.S., as defined by the Census of Manufacturers of 1929 (1). By analyzing rates of variations in these data, he arrives at a temporal description of the deviations of areas in relation to national averages, and the regional displacements of activities over a period of sixty years. McLaughlin's work appears to be the spatial analog of the methods of analysis of fluctuations practised in the U.S., between the wars, by authors such as Burns and Mitchell. Meanwhile, his attempt to provide a new explanation necessitated the transition of dynamics. The differential rates of growth are interpreted as the result of regional differences, on the one hand in the cost of production and on the other in the comparative economic maturity of the different sections of the country. A similar method is still used in the U.S., for example, by the Department of Commerce and the National Industrial Conference Board.

The great crisis of 1929, which was responsible for irreversible locational shifts of economic activities, was to pose for some capitalist countries the problem of regional economic development. In the United States, the policies of the New Deal included in their accomplishments the famous experiment of regional planning in the Tennessee Valley; in France "Compagnie Nationale du Rhone" was established; and in England, the contrast between the development of the London region and the backwardness of the former industrial regions of the North was to result in the creation in 1937 of the Barlow Commission to study the problems of the distribution of industrial population. From 1934 on, by virtue of the "Special Areas Act", and especially after 1937, the British government intervened to facilitate necessary local adjustments. Universities started regional research to differentiate between cyclical and structural unemployment. From this combination there came, in 1939, Dennison's important contribution to the study of depressed areas [108].

For the preceding ten years, urban concentration, depressed areas, and defense considerations, had attracted general attention to location problems. Dennison used the methods of the "Economics Research Department" of the University of Manchester. The empirical part of his work, which is superior to the theoretical part, explains the local surplus of unemployement as compared to the national average. Several statistical series on population, employment, net product, and births and deaths of firms describe "the movement towards the South". Finally, the study reveals the experimental character of governmental policy. Lacking sufficient funds the government tried to help by encouraging transfers. In particular, it favored relatively the prosperous areas, by furnishing them with adequate supplies of labor. In this manner, liberalism faced the dilemma of growing regional inequality or increased governmental intervention which went against its basic philosophy.

In the theoretical section at the beginning of Dennison's work, the author is disappointed by Weber's theory (2). He criticizes Weber for not having selected his location factors inductively, in terms similar to those of the "empty boxes" controversy, also for having shifted his analysis from economics to the technical, and finally for postulating the immobility of the location of labor. In his view, previous contributions were always inclined towards general theories which provided explanations based on a series of formal principles, while empirical studies were limited to actual description of reality. However, his method of approach remained greatly influenced by Weber. He starts with the components of cost per unit of product; the percentage of each component in the total cost measures the importance of a variation in that cost caused by a change in location. This calculation in terms of cost, while it is true that it does not postulate a perfect optimality of location patterns, does not imply either that these patterns are arbitrary. It is especially the investigation of total development which leads Dennison to classify industries according to the relationship between the type of enterprise and its location and to separate the explanatory location factors, as did Weber, into two groups: transportation and other factors.

The analysis of depressed areas is thus related to the problem of location and is structured as part of a study of the irreversible processes which cause major shifts of economic activities.

In the 19th century, the main industries used heavy raw materials which are bulky and have low value per unit of weight. Transportation constituted an important factor of location and industries were attracted to the sources of raw materials. After the First World War, on the other hand, processes were developed using inputs whose value per unit weight was increasing as compared to their cost of transportation. In addition, electricity freed these processes from their dependence on the coal mines. The general use of c.i.f. pricing brought industries closer to their markets and the introduction of the automobile also affected the location of expanding industries.

While the attraction of raw material sites declined, that of market locations rose. This is due to a variety of causes, the effects of which are cumulative and independent of the cost of transportation. They include contact with the consumer, time saving, flexibility of the highway transport system, diversification of products, consumer demands, the satellite character of certain activities, the foresight of entrepreneurs, and customs regulations which tend to protect new industries.

Dennison first isolates labor from his second group of location factors. Weber had reasoned mainly within the framework of the firm and had postulated a perfectly elastic supply of labor at given and fixed centers. Dennison, on the contrary, integrates the equilibrating labor migration in his analysis and studies the secular and total evolution of locations. The strength of the attraction of labor centers appears weak, due as much to the existence of variations in the quality of labor required as to the development of labor unions, which tend to repel firms from those areas where they are strong.

Finally, Dennison introduces numerous factors whose spatial effect appears to him to be variable: the weak effect of fiscal charges, the important but derived effect of capital, the growing effect of construction and free spaces, the growing effects of the conditions of hygiene, publicity, low rent of land, and available buildings. The importance of these factors grows in an economy of imperfect competition and monopoly. The attraction of the markets arouses a cumulative local expansion in prosperous areas while the specialization of the abandoned coal regions provides no resistance against depression.

Dennison identifies a composite force whose elements should be weighted, as opposed to the unique force of agglomeration found in Weber's theory. He returns, in this way, to Marshall's notion of an "industrial atmosphere". The spatial problem is looked upon as a case of the particular clustering of the factors of location and of the adaptation of areas to changes in conditions.

The social and economic problems posed by human concentration and depressed areas, respectively, induce the author to examine the conditions and the limits of intervention of the State in matters of location. Dennison distinguishes two kinds of remedial action: encouragement of industry to remain in the depressed areas and movement of workers to prosperous areas. Whichever of the two methods is chosen, there is still a choice as to whether a liberal policy (of subsidies and licences) will be allowed or whether comprehensive planning will be adopted. Both alternatives pose problems of definition, such as optimal concentration, depressed area, industries to be developed, duration of the intervention, etc., but the difficulty of their application and the risks involved vary as much as their results. Dennison favors a policy of flexible control, less by virtue of objective criteria than by reason of his doctrinal preference for liberalism.

One of Dennison's merits resides in his consideration of the intervention of the State in decisions which, up to then, had been considered the sole province of private enterprise. This problem was not thoroughly discussed again until Hoover's work published in 1948. Losch was to make only a few ancillary contributions to this problem, since State intervention was likely to distort his abstract model of optimal location.

---

(1) McLaughlin (G.E.), Growth of American Manufacturing Areas, Pittsburgh: Bureau of Business Research, 1938.
(2) Cf. Robinson (A.) in "Book Reviews", Economic Journal, June-September, 1940, p.266ff.

Chapter 8. August Losch

INTRODUCTION. THE BACKGROUND FOR LOSCH'S THEORIES

Losch's work, although later than Palander's, is still contemporary from two points of view. Historically, Losch's ideas were made progressively more precise at the same time; and the first publication in 1940 of his main work, "Die raumliche Ordnung der Wirtschaft", constitutes the culmination of these thoughts [112]. Logically, Losch sought in his research the same fundamental objective as Palander, that is, the elaboration of a general equilibrium theory of location in an economic system. But, whereas Palander did not achieve a general theory, Losch made decisive progress toward developing a general model of spatial interdependence.

It is convenient to look back at the main directions of Losch's scientific thought in order to understand fully the content and especially the structure of his theory. Losch's development of the problem appears, indeed, to be closely tied to the development of Losch himself. He starts by studying the connections between demography and economics during an era that is preoccupied by controversies over theories of stagnation. The fact that these first attempts do not lead Losch away from classical theory does not refute their influence on his philosophy. In the analysis of the connections between demography and economics, he perceives the possibility of developing a dynamic economic theory and, in spatial analysis, of connecting space and the economy; that is, the distribution of economic activities in space. In other words, the dynamic approach must take into account the economic development and the spatial approach must take into account the particular spatial characteristics of the economic system. This does not mean that spatial analysis must be reduced to static theory, since the dynamics of the particular spatial characteristics of the economy are an essential element in the resulting economic system.

Losch became interested in the theory of location because of his interest and research in international trade. Convinced of the insufficiency of the law of comparative advantage as a basis for the theory of international trade, he substituted an analysis of the location of economic activities for the classical theory [68] [105] [109] [121]. The controversy which was to take place, mainly with Fritz Meyer, emphasizes Losch's break with traditional thought [120] [138]. In particular, Losch had to make more precise his own theory by the solutions he gave to the problems of "transfer" (Transferproblem) and "combination" (Kombinationsproblem).

After being interested by the theory of location, Losch then tries to generalize it. The publications of Predohl and Palander had made him conscious of the need for a general theory of spatial equilibrium. Knowing the content and the limitations of all the previous contributions, he engages in the formulation of a system of equations sufficiently complete to describe spatial interrelationships of an interdependent economic system. His system is an enlargement, but in the tradition of the work of Walras on non-spatial equilibrium. He limits himself to a compilation of the equations and unknowns by which the abstract interrelationships of a set of locations in an economic system can be described. However, he does not further develop this system of equations, which he thought was too general. Instead he sets up a theory of regions. This is conceived as being the logical and necessary link between the theory of individual location and the theory of general spatial equilibrium. It is in the search for such an intermediate theory of regions (Wirtschaftsgebiete) that Losch makes one of his major contributions to spatial economics [104] (1).

Although the theories of "transfers" and regions doubtless constitute his two most original contributions [247] his major work, to which all previous steps lead, provides greater potential insights [112][259]. Losch's ambition is to re-integrate, within the general theory of location, not only his own theoretical schemes, but even more those of his predecessors. This leads him to elaborate a conceptual framework broad enough to organize the results previously obtained and to integrate his new contributions. The importance given to the theories of "transfers" and regions in Losch's conception appears to determine the structure of the whole system with which he culminates his work. His book starts with a systematic examination of the problems of location (Standort). He then poses the fundamental conditions of the general spatial equilibrium. However, the predominance given to the theory of regions encumbers Losch's general approach. Ultimately his regional theory becomes the center of his later theoretical development; but it appears to be merely coexistent with the independent theory of location which could not be integrated with it without seriously modifying the formal framework. The connexion of the theory of "transfers" and that of regions in Losch's mind, gives an explanation to the fact that the theory of trade, in the largest sense (Handel), is better related to the theory of regions than the theory of regions is to the theory of location. It is logical, therefore, to construct the theory of trade on that of regions: that is, to join it to a theory of interregional relations, since trade theory was henceforth to be founded on a spatial division of labor.

The development of Losch's theory reflects the particular steps taken by the author, which also provide the explanation for its insufficiencies and its limitations.

Section 1. THE THEORY OF LOCATION

Following Palander's approach, Losch tries initially to formulate a systematic treatment of the problems of location. Then he adapts his general conclusions to the principal types of locations (agricultural and industrial locations, concentration in the cities and belts). Because of this, various levels of observation are mixed, so that Losch's model had a very distinctive character. His theory appears to be the logical and necessary result of the previous contributions carried to the highest level of abstraction and generality.

Losch admits that, while existence in time is determined, location in space can be selected, even though this choice is influenced by the place of origin. The possibility of choice calls for a hierarchical arrangement of knowledge and the author distinguishes between an actual location and a rational location. An analysis of actual location patterns, either historically or based on current behavior, leads to verification and explanation of the considerations that guide the entrepreneurs in their choice of location; the theory of rational location tries, on the other hand, to determine ideal location patterns abstractly. These two models do not coincide and Losch tries to measure the deviation that separates these two types of location patterns. The deviation is found to be smaller than would have been expected.

The ideal location for an individual entrepreneur, differs from his ideal location as seen by society, so location must be defined successively at the level of partial and general equilibrium. For partial equilibrium, all locations are given except one, which is to be determined by maximizing the relative advantage of the remaining locator. If a production site is sought, the locations of the factors of production, of the competitors, and of the consumers are given; only their relations vary: the production of a farm takes place on a surface and is dispersed over some points, while that of an industrial firm takes place at a point and is dispersed over a surface. The location of the consumer depends on the neighbouring consumption centers and on production location. Two sets of impacts occur which connect the partial and the general equilibria.

The location of the firm has, first of all, an influence on its competitors; Losch returns, therefore, to the problem discussed by Hotelling, even though he refuses to generalize about the tendency toward concentration of firms. The location of the firm also influences its customers and suppliers by affecting the structure of their activities. The type of consumption varies with distance (a piece of machinery, for example, will be delivered assembled up to a critical distance, beyond which it will be assembled at the destination). Similarly, the intensity of a process of production is a function of the distance from the point of consumption, since local manufacturing becomes more expensive than central manufacturing where agglomeration economies occur. Consumption, on the other hand, varies with the distance from the point of production especially when several kinds of merchandise enter into competition. Conversely, changes in the goods produced caused by distance from the point of consumption brings Losch back to Thunen's model of concentric rings.

Meanwhile, each location has impacts on all the others. These repercussions lead to a general interdependence. The basic locational force leads to a pattern which tends to equalize advantages of the individual units and to maximize the number of competitors. General equilibrium, in order to be expressed by a system of soluble equations, must take into account the conditions of the entire system. These conditions go well beyond what the partial theories of location can determine.

Losch is thus led to insert the theory of the economic region between the partial and general theories of location in order to show the interdependence of the locations themselves.

The result of choosing an ideal location leads to characteristic combinations of production and consumption points, each group of which constitutes only a part of the total market. While the theory of location had previously rested on the hypothesis that there is a one to one correspondence between a place of production and a place of consumption, Losch's theory goes on to consider the case of multiple centers of supply or demand. Two basic types of positional relationships, independent of the number of units of production or consumption located in market areas, constitute the basis for every location choice. While the areas of demand play the most important role in the theory of industrial location, it is the areas of supply that dominate the theory of agricultural location. It is not the number of units, but rather the number and the positions of their locations, that determines the nature of a region. These locations can be scattered or concentrated. Losch distinguishes between two basic types of concentrations: punctiform agglomerations and area agglomerations. The grouping at a point which is the optimal location for numerous producers leads to homogeneous or heterogeneous point agglomerations, depending on whether production is of similar or different commodities. The market surfaces for such points tend to coincide. In area agglomerations, however, they tend to be adjoined. The area agglomerations of homogeneous locations become, in Loschian terminology, "belts" (Gurtel) when their market networks are grouped, or "districts" when only their centers are grouped while their markets are separated. Finally, the area agglomerations having heterogeneous locations lead to industrial regions that partly resemble the belts and partly the districts.

Losch's approach compels a definition of both the boundaries that separate the simple regions and those that separate systems of regions.

The boundaries of the simple regions separate the market or supply areas of heterogeneous products (for example, Thunen's rings), or they separate regions which sell or buy homogeneous products in different localities. While the boundaries between the areas of competing products are given by their locations, the boundaries between competing locations must be calculated from these locations.

The systems of regions are distinguished by either a differentiation of regions in the production of goods (Thunen's system of rings) or by the overlapping of regions ("economic provinces"). Their limits are either boundary lines or boundary zones. The boundary lines mark the boundaries between the simple regions, and separate market regions for the same product when two cities or two provinces have similar structures. If the size of the towns or provinces is unequal, on the other hand, the product on each side of the border differs in quality and quantity. While the lines trace a clear boundary between the regions of two cities or two capitals, the boundary zones determine a border market region between the two capitals which does not belong exclusively to either district. Being located on the edge of the market network, these markets are too small for a new entrepreneur to become established and cannot acquire sufficient size by merging with similar zones of the neighboring province.

However, these problems become more or less important when they are examined in the context of the typical problem of location.

Consider first industrial location. The economic calculation of the entrepreneur leads him to choose a location where his profits will be maximized. The theory establishes an approximate formula which depends on a number of rational factors which are chosen among the entire possible set. It would be neither scientifically possible, nor practically expedient to take all factors into account. Elaborated in terms of nominal profits, (the difference between cost and receipts), Losch's theory goes beyond the previous analyses which located the firm at the point of minimum delivered costs, presupposing, as did Weber, an inelastic demand. Thus the author progresses from the simple consideration of one set of factors affecting location, such as expenditures or receipts, to a joint consideration of the elements which determine profits.

From the point of view of costs, the location of the firm depends on the cost of transportation and production and on the relationship between them. The search for the point of minimum transport costs brings Losch back to the studies of Launhardt, Weber, Palander and Predohl. At times, costs alone suffice to orient production; this is always the case where production is attracted to an extreme location, such as the place where the factors of production, the consumers, or the competitors are located. This is not true when there is a technical orientation. In this case, location remains indeterminate, except where a fixed factor or a fixed combination of factors is used. However, the costs of production only influence location through the costs of transportation, for as a rule, production costs cannot be separated from the freight costs of raw materials. In the last analysis, the cost of transport decides the choice of the source of the raw material, for example, and the cost of production determines which of the point of lowest freight rates will be chosen. Thus location depends upon total costs. Again, the choice of an extreme location can be explained by total costs when the advantages in terms of production costs or transportation costs are not determinant.

However, the location of the best market, where revenues are highest, may attract the firm. Orientation to the location of the best market can be further subdivided into orientation according to either one of its components: quantity or price. An orientation to quantity would consider mainly the number of buyers, while an orientation to price would consider purchasing power. When the market is technically tied to the buyer and the location of the clientele is fixed, consideration of gross receipts alone forces production to the location of the consumer.

All the points of view so far discussed may perhaps explain an actual location. But generally, these factors alone cannot indicate the ideal location which maximizes net profit. The location of the greatest profit coincides with that of the lowest delivered costs only where revenues are fixed. But demand and prices are not constant, as Weber assumed. Losch underlines the close interdependence that relates price with demand and location. However, he forgoes an exact determination of the rational location. A geometrical solution becomes impossible as the optimal orientation involves more than three variables. Algebraic treatment leads to insoluble equations, since there is more than one point where the total demand can be at a maximum. Between these points, demand is not a simple function. For this reason the total possible demand must be determined separately for each of a number of possible factory locations; and for similar reasons the best volume of production must be determined as a function of factory price. The empirical solution requires a return to market and cost analysis. If complete orientation is considered, production necessarily takes place at an extreme point only when it depends on an irreplaceable and immobile factor, consumer, or a combination of both. In general, in the theory of industrial location, the term "orientation" can be used to designate either the motives for or the outcome of a choice of location. With respect to an "extreme location", orientation can be definite or indefinite, according to whether only one or more than one site could provide the necessary conditions for location. Losch, after studying the relationship existing between the different types of orientation and the cases of extreme locations, concludes that no single factor can indicate the optimal location, with the sole exception of an unavoidable technical restriction at one single source of supply.

Turning next to the theory of agricultural location, Losch considers the problem of location from the standpoint of the optimal spatial arrangement of crops, rather than from the standpoint of the individual unit and the maximization of its profit. He tries to generalize Thunen's theory, starting from more general assumptions. Thunen's famous rings represent only special cases of a more universal theory. In Losch's general model, the spatial ordering of the crops can vary widely. Losch studies mathematically the conditions which the production cost per acre area, the physical yields per acre, and the market prices per unit of weight of two agricultural products must fulfill in order for one of them to be cultivated at the center and the other at the periphery of a given area (2). These conditions of course require that the rent per acre provided by the first of these products must be greater than that of the second. In addition, certain restrictive conditions must be met by the relationships of the costs, yields, and prices of the two products in order to guarantee the production of the second product beyond the center ring. Of the 27 cases studied, Losch finds that only in ten was the choice of a crop related to distance. It is only in these ten cases that the concentric rings may appear. In fourteen cases monoculture is practised, one of the products being excluded from cultivation. In three cases, the two products are simply cultivated side by side, their location no longer following an ordering principle. The system of equations developed by Losch allows two unknowns to be determined simultaneously: the volume and the location of cultivation. Here, location no longer depends on specific value and weight of the product as in Thunen's model, but on the physical yield per unit of cultivated surface. Moreover, not only do Thunen's concentric rings appear here as a special case, but in some situations they can even be inverted: that is, they can lead to an inverse ordering of the products. It is sufficient to modify one single assumption for the inversion to occur; if the product that would have located in the center is imported instead, this product will be forced to locate on the periphery.

Losch proves geometrically that an equilibrium situation can arise in a traditional economy and that Thunen's rings must be formed in a dynamic economy. Finally, with respect to the comprehensive solution, Losch discusses the location of agricultural production taking joint production into consideration. Every enlargement of a sector takes place to the detriment of another, equilibrium being reached only when the physical yield per unit of surface is the same for all locations or when the costs of transport are inversely proportional to the yields per unit of surface.

The dichotomy between industrial and agricultural location which is at the foundation of Losch's theory leads him finally to note that the number of locations of production is larger in agriculture than in industry, but that the number of locations of consumption is smaller. The market for agricultural commodities is punctiform and results in free competition, while the market for industrial commodities is areal, resulting in limited competition. But in both cases, the maximization of the number of producers is assured, and the size of the units is determined both by the factors that are peculiar to each type of location and by those that are common to all.

Losch goes from the analysis of individual locations to that of the agglomeration of locations. He studies the reasons for town locations, which are regarded as point agglomerations of non-agricultural locations, and the location and cause of "belt" formations, which are concentrations of agricultural locations in areas.

With regard to the towns, Losch distinguishes four natural causes that could explain their formation. First the advantages of large scale production may account for the concentration at one point of large individual enterprises. Similarly a town can be born from the agglomeration of similar enterprises attracted to the same place by the advantage of numbers and the external economies they may derive, by the advantages of site and source of supply, or by the fact that competition is not possible from outside this market area but is possible inside. On the other hand, the tendency toward the agglomeration of sellers, studied by Hotelling and based on freight cost, depends on hypotheses that are too restrictive. Losch claims that with any degree of realism such a tendency would not hold. Agglomeration of heterogeneous enterprises can also create a town, whether this concentration is due to interdependence or chance. Finally, towns may be started by a simple agglomeration of consumers, which results for the analogous reasons as agglomerations of producers. But to these natural causes of the birth of towns, Losch adds historical factors and tradition, by virtue of which locations tend to agglomerate around a pre-existing source of supply. However, all these causes cannot, by themselves, account for the location of a town. For, if the locations of all towns except one are considered as given, it is impossible to determine the location of the latter by searching for the point of coincidence of the location of individual firms, since the interdependence of the locations of the various units concerned and the non-economic factors of their concentration cannot be disregarded. This methodological difficulty leads Losch only to outline the possible hypothetical situations (for example, the points of intersection of various lines of communication). He also perceives, as in the case of the firm, the possibility of uniquely advantageous situations such as natural resource sites, although their importance should not be overestimated. On the other hand, the system of town locations may be rigorously determined, but its analysis depends on the theory of economic regions.

The same reasons that explain the formation of towns - advantages of position, site and concentration - account for the formation of "belts". This is true for areas producing identical goods which must find an equilibrium in a compromise between specialization and diversification, or for areas producing heterogeneous goods, such as industrial districts.

Finally, Losch's theory of location is enlarged into a general spatial equilibrium model. It may seem surprising that Losch would treat such a model as part of a study of specific problems of location (theories of industrial and agricultural location and of agglomerations and belts). In generalizing his theory, he is led necessarily to a higher level of abstraction in order to achieve a systematic presentation of spatial problems. In this treatment can be seen a consequence of Losch's position with respect to the theory of general spatial equilibrium. He is in fact convinced that a free economy functions rationally and satisfies the conditions of general equilibrium. Therefore, he is able to perceive the major theoretical gains to be derived from the integration of the spatial factor into Walras's equations. He is even able to see some practicable application of a general model, since he attributes the failure to reach the optimal equilibrium to human intervention. His new model provides a needed insight into the real conditions of the system. However, he is too convinced of the essentially abstract character of a model of general interdependence to give it a prominent place in his analysis. This is why he contents himself with simply stating the main equations that express the general equilibrium conditions without developing them. As a result he assigns his model only a secondary place in his work (3).

Equilibrium is determined, according to Losch, by two fundamental tendencies: the maximization of individual advantages and the maximization of the number of independent economic units. At the equilibrium of these two tendencies, the model of interdependence determines the locations. The model can be formulated as a system of equations expressing five fundamental conditions. First, the location of every individual unit must be advantageous as possible. Second, locations must be numerous enough to occupy the entire space. Third, it follows that abnormal profits must disappear. Fourth, the areas of supply, production, and sales must be as small as possible: that is, of a size just sufficient for the survival of the individual units. Finally, with respect to the boundaries of economic areas, it must be a matter of indifference to the individual units to which of the two neighboring areas they may belong. These boundaries are indifference lines. But Losch shows that the satisfaction of all five conditions does not guarantee that the best locations for production and consumption will coincide. On the contrary, the fundamental dichotomy between agriculture and industry must be remembered: that is, that the optimal location for the production and consumption of industrial goods is located in a larger city, while the optimal locations for the production and consumption of agricultural goods have a uniform distribution.

Losch does not proceed beyond making an inventory of the unknowns and the equations of the general spatial equilibrium system. Later works, such as those of Isard and Miksch, are analytic studies of the conditions of the equilibrium and explicit penetrations of its functional forms. Losch concentrates on elaboration of a theory of economic regions. He conceives this as the necessary intermediary link between the theory of general interdependence and the theory of location of individual units.

Section 2. THE THEORY OF REGIONS

Contrary to the postulates accepted in theories of international trade in which political and economic boundaries coincide, Losch tries to delimit purely economic regions (Wirtschaftsgebiete) and to show how economic boundaries can be defined. As a result of his concern for relating location, region, and interregional and international trade, Losch reconsiders the work of Weber and Ohlin.

Following a procedure that has since become famous, he starts from extreme assumptions which allow him to eliminate all spatial differences of non-economic origin so as to approach an ideal model of the region by analyzing only what is essential. He assumes an extensive uniform plane, where raw materials are evenly distributed. This surface contains no political or geographical inequalities and supports self-sufficient farms which are uniformly distributed. Spatial differences will be created, however, from the interplay of economic forces. The advantages of specialization and mass production will lead to concentration, while those of diversification and economies of shipping costs will favor dispersion.

If one of the farms were to produce more of a product-beer, for example - than it needs, it would benefit from economies of scale; but the cost of shipping could limit the distribution of its product. Losch, basing his reasoning on the classical assumption of a negatively sloped demand curve, shows that since shipping costs increase with distance, after a certain limit an extreme sales radius is reached. The relationship between the cost curve and the demand curve determines the length of this radius; and the market area, given the assumptions of uniformity, is a circle (4).

If the breweries are so numerous that **their** sales areas become contiguous, the market shape would cease to be circular, since the circles would have empty spaces between them. Following Chamberlin's analysis, new entrepreneurs would be able to enter the market, thus reducing the size of the individual firms without making them unprofitable. A network of hexagons is substituted for the circles, with the hexagons being smaller than the extreme sales radius circles that circumscribe them. The demand curve is slightly displaced to the left and the size of each hexagon is reduced until the corresponding demand curve becomes tangent to the supply curve. The market is then fully covered. The appearance of new sellers would make each sales area too small. Just as there exists a radius of the most extreme sales area possible there also exists a minimum radius of the sales area necessary to support production.

Unlike the square or the triangle, which are the other two possibilities for eliminating the spaces between market circles, the hexagon represents the geometric figure nearest the ideal circular form, so that the largest demand per unit area can be obtained equally by all producers. It is, then, the optimal market area shape.

Proceeding from the assumption of a continuous distribution of the population to the case of discontinuous distribution, where the population is grouped at uniformly distributed points, Losch shows that the hexagonal model is still the optimum. According to the orientation of the hexagons, a center of production will supply a variable number of settlements. Three typical arrangements of hexagons appear with settlements located at the vertices of the hexagon, at the mid-points of the sides or entirely within the hexagon, the sides of the hexagons lying in the empty space between settlements without touching them. In each case the necessary shipping distance between the seller and the farthest buyer varies. This distance is the maximum at which a commodity can be sold if its production is to be profitable. Losch shows graphically that each seller satisfies the equivalent of the demand of three settlements under the first typical arrangement, four under the second, and seven under the third, taking into account shares he might have to concede to the competitors.

A center of production can have a market area of various sizes, depending on the relationship between the supply and demand curve and shipping costs. Losch superimposes on the same diagram ten hexagons alternately set out following the three typical arrangements and shows how the location of competing centers is determined when the first center has one or another of the ten sales areas. As the necessary shipping distance for the market surface considered is known, it is only necessary to establish competing centers at a distance of twice the market radius from the origin center. These points become the centers of six identical hexagons surrounding the first and each equal in size to it. The same operation can be repeated around the new hexagons and so on. This graphical procedure demonstrates the existence of sectors dense in production centers or "city-rich sectors" (die stadtreichen Sektoren), a questionable expression that shows that Losch insufficiently distinguishes between homogeneous and heterogeneous concentrations of locations.

Finally, the location of the center of production itself can coincide with a settlement or be located between three settlements. In this last case, it defines the location of a center of gravity and while the center of production has no local demand, it is nearer to the other sales points. Theoretically, the firm locates at that location which provides the largest total demand for a given number of places. But, historically, and for practical reasons, location at a settlement is more frequent.

It is in this manner that the network of commercial areas appears, from the smallest to the largest according to the products involved. The problem of the respective locations of the centers of production of different types of goods in relation to each other leads to a consideration of the system of networks.

Losch points out that it is possible to arrange these networks randomly over the surface under consideration. In spite of the resulting chaos, every place would lie in a market area for every good. But several considerations suggest a more ordered and more economic arrangement. Losch starts by orienting the networks so that all of them have at least one center in common: a metropolis. This point benefits from all of the advantages of a large local demand. If the networks are rotated around this center in such a way as to get six sectors with many and six sectors with only a few production sites, not only will the resulting arrangement not deprive any of the places of their access to all products, but it will also provide optimal lines of transportation. This is true since the more the centers of production coincide, the greater the number of purchases that can be made locally. On the one hand, shipping distance diminishes; on the other, the length of the transport routes is reduced and the shipment cost is minimized. Losch finally reaches a complicated but ordered system. The network of market areas and the number of self sufficient systems depends on the product whose necessary shipping radius is the largest, taking into account the economic limitations in the size of the central city.

In the case of a discontinuous distribution of the population, only those networks of markets lying obliquely can be rotated. If a circle is traced around the center of production, having as radius the diameter of the market area that possesses the third type arrangement, this circle passes, in its first quadrant, through three settlements. These give three possible locations for the center of the nearby third type areas and there are two possibilities for the rotation of the networks. Of the three centers lying on the quadrant, either of the two outer ones, which are at the same distance from each other as from the original center, may be chosen as centers of two of the neighboring hexagons. Otherwise, the middle settlement is chosen as a center, together with a fourth point at a proper distance in the next quadrant. Depending on the choice, the middle sector or the two outer sectors will be relatively poor in cities. These "central places" of Losch, who uses Christaller's terminology [76], are regularly distributed. It is true that towns of the same size may quite possibly fulfill entirely different economic functions. A smaller agglomeration generally lies about halfway between two larger ones and the size of the agglomerations increases with their distance from the metropolis. The description of the ideal landscape includes the location of the lines of communication that unite the centers of the regions, and whose relative importance varies as a function of the number of centers. The number of centers of production per unit of length can serve to measure the density of traffic.

Losch's model includes the whole range of market areas, even though they may not all exist in reality, either because some market area may not coincide with the necessary market area of any product or because of extra-economic factors that influence spatial economic arrangement. Administrative divisions are frequently carried out in such a way that a certain number of smaller regions are combined into a larger district.

The identity of the structure of the market areas that Losch finds leads him to a particular model of systems of areas, each of which includes several others of the next smaller size. The spatial picture is then simplified, but this is achieved by some sacrifice of economy since many goods lose their optimal sales area.

Finally, if square market areas are substituted for hexagonal ones, the systems of networks are still formed in the same way. City-rich and city-poor sectors appear once more. However, the lines of communication in a square system are longer, so that the hexagonal system remains the only one that is economically optimal.

The networks of market areas for heterogeneous goods, systematically arranged, define the "economic region" in a strict sense.

On the other hand, the relationship between similar spatial systems leads to the formation of a new network, the network of systems, which locates the set of systems by reference to each other. Each system is no smaller than the largest simple market area that it contains, nor is it too much larger. If this were not so, it would mean either abandoning arbitrarily the production of goods whose minimum necessary shipping distance is larger than the radius of the system, or that all the places located at a distance greater than the radius of the system would have to forego the establishment of a second metropolis. The advantages of the agglomeration of locations at the center of gravity of a system are not noticeable in the distant regions which are not supplied by the center. Similarly, with distance, the difference between the city-rich and the city-poor sectors and the advantages of communications associated with it disappear progressively. Hence a second metropolis arises at a distance from the first one equal to twice the radius of the system of networks and so on, until finally the entire plane is filled with systems. When these have become numerous enough they assume the shape of regular hexagons. But it is still undecided how city-rich and city-poor sectors will be oriented with respect to one another. In effect, in accordance with their disposition, the lines of communication will go through territories that are the richest or poorest in cities. All industries find at least one location in every system. Only those goods with a necessary shipping distance that is not much greater than half of the radius of the system are produced in more than one place. For those whose shipping distance is less than one third of the radius of the system, it is impossible that the network of their market regions should end at exactly the boundary of the landscape so that there will be empty corners. Finally, those whose shipping distance lies between one third and one half of the radius of the system will have their locations on or near the boundary of the system and their market areas will generally encroach upon the neighboring system.

This division of the regions into such networks completes Losch's theory of economic regions. In it he explains the organic spatial structure by purely economic considerations starting from a primitive uniform plane.

However, this model, derived simply as a function of three factors, distance, mass production, and competition, is then complicated by the introduction of a series of important elements that Losch divides into four groups, each introducing a type of differentiation that distorts the simple model. These four are economic, natural, human and political factors.

Economic distorsions can result from local differences in prices, products, or transportation rates. The study of local differences in prices leads Losch to integrate spatial theory with the theory of monopolistic competition. After making the elasticity of demand a function of distance, he is able to determine mathematically the prices and the general equilibrium conditions of the market (5). These equations enable him to solve the problem of choice, confronted by the seller, between three ways of fixing prices: (case A) to establish a different price for each individual buyer, (case F) to keep prices so rigidly fixed that all buyers pay the same f.o.b.price, or (case C) to adjust prices so that all buyers pay a uniform delivered price. If linear demand is assumed for each individual buyer, the elasticity of this demand appears to be greater in the case of C than in the cases of A and F. The same is true for a supply area surrounding a purchasing center. Every variation of the price in the center affects the producer who is far from the center more than the producer who is near, since receipts gained by the farthest seller are proportionally smaller and thus his supply is more elastic with respect to uniform delivered prices than f.o.b.prices. By reconsidering the comparison between industry (market area) and agriculture (supply area), Losch concludes that, in the industrial economy, demand is more elastic than supply and that the inverse is true in an agricultural economy. Consequently in a general way, the individual demand under uniform delivered pricing is more elastic as the distance becomes larger, compared to f.o.b.pricing. Losch then proceeds from a consideration of individual demand to that of total demand by equalizing the demand of a number of scattered locations to the number of individual demands of a single locality.

From this he finds that in case C the demand of the market area is smaller and more elastic than if the same individual demanders were concentrated in one point. In the case of A and F, the demand at each point is less than that of the central location. Under these conditions, the seller will practice a price policy such that each buyer pays the f.o.b. price that gives the seller the maximum profit. Case A is preferable from his point of view to the cases of F and C; that is, complete price discrimination is preferable to price uniformity. Losch's and Hoover's equations lead to the establishment of a formula for the f.o.b. price that is optimal with respect to the location of all the clients, since it ties the elasticity of demand to marginal costs and the cost of transport.

However, spatial price discrimination tends to be self-limiting. As the elasticity of demand increases with distance, the f.o.b.price quoted to distant buyers must diminish. But price discrimination raises the marginal costs as it reduces the market area, and consequently the profit-maximizing f.o.b. price also tends to rise, so that the extent of discrimination tends to be automatically limited. Finally, the possibility that the farthest buyers may buy from other buyers closer to the place of production prevents spatial price differentiation from exceeding the difference in transportation costs. Important applications of spatial price discrimination can then be analyzed for cases of dumping, overlapping areas, and "basing points". Losch distinguishes between the tendency to maximize individual profits and the tendency to maximize the number of sellers. The general market equilibrium may be achieved either in a system of open competition (Chamberlin's free entry) or in a system of closed competition. The theory of market networks had already shown that the minimum size of the individual area is determined, when new firms enter the market, by the tangency point between the demand curve in terms of uniform f.o.b.price and the marginal cost curve.

Under these minimum-market-zero-profit circumstances, any additional reduction of the market area implies a discriminatory price policy and the pressure of competition leads to price differentiation. If competition takes place, on the other hand, in a closed market, the policies of price differentiation must adjust themselves so that competition may be maintained at the limits of the total area and so that in the interior of such an area a stable equilibrium position is reached. Moreover, the locations of the sellers and buyers react differently according to the way in which the prevailing price is determined. In the case of free entry and spatial differentiation of prices (case A), the locations of the sellers tend to concentrate. Cases A and C benefit distant consumers more than nearby consumers, while with "basing point" pricing the locations close to the base point tend to benefit. For the buyer, the effects of these pricing systems are different. Buyers are attracted toward the place of production in cases A and F. Their location is indifferent in the case C; and the "basing point" policy attracts them toward the base point. For given locations, case A appears to be the most favorable to the producer while case F is, on the average, the most advantageous for the consumers. Finally, proceeding from the analysis of price discrimination within one market area to that of price discrimination among different market areas, Losch presents a description of different geometric form representing the market areas according to the relationship between the prices and the cost of transportation. Thus, the schemes developed by Launhardt, Palander and others are integrated into the analysis as polygonal patterns that result from the distortions of Losch's regular hexagons.

Although spatial price discrimination reduces the size of market areas, product differentiation counteracts this effect, whether the product differentiation is a function of the place of production or not. Since the demand for a product does not come exclusively from the immediate neighborhood of its production site, the boundaries of its sales area no longer coincide with those of neighboring production sites, so market areas can overlap. Product differentiation appears, then, as a way of penetrating into a competing area, by means of the inelasticity of demand. Nevertheless, this product differentiation depends itself on the ease of product substitution, the product's price as compared with the cost of transport, and on its relative uniqueness.

Finally, Losch points out a last economic cause of the distortion of his hexagonal model: the local differences in transport rates. He distinguishes between two types of variations in transportation rates. If rates vary with the point of destination, the elasticity of the volume of traffic with respect to the freight rate will rise first, with this rate and secondly, with the distance from the point of destination. Thus, for long hauls, the transport rates are reduced so that transport costs may be less than proportional to distance, but more than proportional to distance for short hauls. The effect of this differentiation on the size of market is different according to whether the differentiated rate is higher or lower than the proportional tariff at the boundary of the minimum necessary sales area. If it is higher, then local prices rise and local demand falls, while these effects are opposite if it is lower. Thus, the differentiated rate lengthens the possible sales distance, but the minimum necessary sales distance is higher or lower than the possible sales distance according to whether this rate is above or below the proportional rate. If rates vary according to the point of origin, the differences in rates make different points of origin unequally competitive. The product from the lower rate origin cannot be excluded from sale, since its c.i.f.price somewhere will fall below that of competitors whose transport rates are higher. Furthermore, its market area surrounds the market areas of the competing products. The latter become all the smaller as their transport rates and/or their f.o.b.prices are higher. Finally, Losch considers the case where the rate variations are calculated on the whole and do not vary with the points of destination or origin. The lowering of the freight rate, by reducing the c.i.f.prices and augmenting demand, enlarges the possible size of the market area, but diminishes its minimum necessary size. Besides, the general lowering of the costs of transport, by raising profits, has a tendency to cause new firms to appear and thus may result in a reduction of individual market areas. To the extent that the increase in the size of the market area is strengthened it will favor concentration. In addition, the reduction of the cost of transport often permits large scale production and specialization, thanks to the increase of the sales area. As a consequence, the lowering of the freight rates tends finally to modify the size and functions of the cities but not their locations.

Losch adds to these economic factors a group of what he calls natural elements, the effect of which is also analyzed in terms of the distortions they introduce into his hexagonal model. To take these elements into consideration, Losch drops the initial assumptions of a uniform fertile plain and transport surface in order to envisage the effects of local differences in productivity and accessibility. Differences in productivity favor differences in human activity according to place. These differences can be randomly distributed, but the disorder that results from the irregular character of their influence is in part corrected by the underlying ordering forces of the Loschian system. If they are, on the other hand, localized at certain places, the distorting force is weakened by the regularity that rules their distribution. If they are distributed uniformly over large distances, they will not give rise either to irregularities or to inequalities. Inside the zone most favored by the natural differences in productivity, Losch notices consequences similar to the effects of a general lowering of transport cost. On the border of this zone, the frontier market areas that extend over the two types of differentiated regions possess unequal parts, the size of each is proportional to the size of the areas in the two zones respectively. Finally, in the case where the two regions are equally endowed, but of unequal size, and separated by a deserted zone, the largest region alone will benefit from the advantages of large scale production.

The substitution of the assumption of differences in accessibility for that of a uniform transport surface leads Losch to study the effects caused by the existence of "lines" and "points of transport". The assumption of lines of transport permit him to integrate into his theory of regions the "law of refraction" of Palander and Stackelberg (6). Losch points out an important limitation of this law: it assumes that locations to be connected are definitely fixed and few in number (two). This means that the law is not applicable to the general Loschian model. The importance of points of transport varies. Those which are equilibrium points in Palander's mechanical model become "nodal" (Knotenpunkte) and are the centers of attraction for numerous locations. Among these points, Losch distinguishes the "bottleneck city" (Torpunkte) which is primarily determined by geographic features or the existence of transhipment points which cannot be bypassed. The more restricted these bottlenecks are, the more important is the market lying behind them and the more they will attract those industries that search for a large market area. In general, nodal points on the transport system also play a market regulating role in that they order and unify the system of prices, while at the same time creating a structure of limited competition.

With respect to what Losch calls human differences, the specific behavior of individuals and groups is integrated into the initial model, which implied their uniformity and perfect rationality. Limited to the behavior of entrepreneurs, the analysis of individual differences establishes a relation between the ability of the entrepreneurs and their production functions which produces inequalities in the size of their markets, their prices, and their locations. With respect to the differences between groups, Losch points to the character of peoples and races and shows how, through their consumption patterns, they exert an influence on the spatial model of the economy.

Political differences constitute the last group of elements that are responsible for distortions that affect Losch's hexagonal model. The intervention of the State in a domain previously regulated only by private decisions poses the problem of the relationship of political territories to economic regions. Losch discusses this problem by describing the similarities and differences between the two types of territories. The similarities include: capitals that control hinterland areas and main traffic routes; locations of capitals are generally central and their dimensions are equal in size, the dimensions of two categories of territories are dependent on the development of transportation and production techniques; existence of irrational boundaries; and decreasing influence of the capitals with their distance to the frontiers. The main differences between the economic and political territories include the greater stability and more precise character of political boundaries and the different hierarchy of needs and goals to be satisfied by the two types of territories. Losch is then able to discuss more precisely the reciprocal influences between economic and political territories. While private interests exert an impact on the policies of the State, the political boundary has an economic impact as an obstacle to mobility. Thus, those market areas that are sufficiently large to be affected by political boundaries are distorted by the double tendency toward the reduction of the number of locations ("no man's land") and their displacement of either side of the boundary. The zone of economic influence of the boundary also depends on the comparative sizes of the two kinds of territories. Finally, in certain extreme cases, the economic territory exactly covers the political territory because it derives its existence from the latter.

After analyzing the four groups of factors that can distort his pure model, Losch retains those that tend to limit the size of the market area. These are the factors whose burden, whether monetary or not, increases with distance: transport costs (the case of increasing freight rates or increasing cost), time costs, selling costs, business risks, product differentiation, growing size of enterprise, commercial boundaries, and organizational factors.

To summarize: Losch's theory of economic regions appears as an analysis of the network of market areas in regional systems, themselves connected in a network of systems. The hexagonal form and characteristic sizes of market areas which arise under simplified conditions, are then complicated using a method of successive approximations. It must be understood that the apparent chaos that results does not exclude an ordering principle.

Section 3. THE THEORY OF EXCHANGE

With the help of three main factors (location, supply, and large scale production), Losch explains the causes of the division of labor. In his theory of exchange (Handel), he considers the nature of this division, while trying to integrate the theory of exchange with that of economic regions. To the extent that Losch follows the work of Weber and Ohlin, he is led to elaborate a theory of exchange in the widest sense of the spatial division of labor. He is then able to develop an interregional trade theory based on the theory of economic regions and show that international trade theory is only a special case of this theory resulting from the disparity of the monetary systems, political boundaries, and customs duties.

Using an iterative method, Losch starts from the three fundamental factors that can explain the nature of the division of labor: the economic subject (man); the economic activity (work or occupation); and the place of location. Combining these in pairs, he poses, following a well known procedure, the "six cardinal problems of the division of labor". Their solutions, if compatible with each other, will satisfy the conditions of equilibrium of the system through the spontaneous interplay of economic mechanisms. Any further complications that may be introduced into the initial scheme are then analyzed as if they were disturbances of the equilibrium conditions. This allows Losch to integrate the "problem of transfers" with that of "combinations" in order to describe the return to equilibrium due respectively to flows of goods and factors.

The description of equilibrium results from the solution of the six fundamental problems posed by the interrelations of persons, occupations, and economic locations. These problems are: the choice of an occupation by the economic agent; the population of an industrial sector; the residential choice of an agent; the population of a locality; the activity of a locality; and the location of an industrial sector. Losch discusses each of these problems in turn.

The analysis of the choice of occupation leads to particularly important and original conclusions. Marginal analysis, of the Walrasian type, leads to the conclusion that if labor can be divided into subgroups that are homogeneous in productivity, the individual members of a particular subgroup are distributed among the various industries in such a way that their wages are everywhere the same, and that the number of individuals of each subgroup employed by each industry is such that their marginal productivity is equal to their wages. Losch diverges from these conclusions, since it would be highly improbable that two individuals would have identical productivities and that any of them could exert an influence on the wage rate. It is in the principle of comparative advantage that Losch discovers the basis of the interpersonal division of labor, after having shown that it is not applicable to the explanation of the division of labor between countries. Actually, except for the very simplified case of two countries and two products, this type of analysis is of little use. In the case of two countries and n products, only the order of the respective comparative advantages of these countries for the production of different goods can be specified. Thus, once the order in which the products must enter into exchange between the two countries is known, the principle of equilibrium in the balance of payments is necessary to determine the dividing line which separates import goods from export goods. In other words, the absolute level of prices must be known, in

addition to the ratio of relative prices, since its variations have the function of producing an automatic equilibrium of payment balances. In the case of n countries and n products, the comparative advantages of different countries taken in pairs can only be ranked for the production of different goods. The principle of automatic equilibrium of the balance of payments, however, allows only for the determination of two things for each pair of countries. On the one hand, it will give the rank of the two products which will determine the point at which the two countries will necessarily become importers with respect to other undetermined countries. On the other hand, for those goods that are included between the two limits of the smallest respective advantages (i.e., the points at which the countries necessarily become importers) the need of each of the two countries to import these goods from other countries is determined. Thus the theory of comparative advantage does not allow the determination of which goods actually are exchanged between two countries. Those products for which each of a pair of countries, considered in isolation, possesses the highest comparative advantage are not necessarily bought by the other country if the number of countries engaged in exchange exceeds two. The products that assure a comparative advantage to a country relative to one other need not end up as export products when all countries are considered. After showing the inability of the theory of comparative advantage to explain the nature of international trade, Losch finds the theory suitable to account for the nature of the interpersonal division of labor and of the choice of occupation. It is sufficient to translate the theory into the proper terms to express the realities of the economy of the individual. After setting up a scale of relative productivity of an individual in different activities, Losch computes the actual number of product units produced per unit of time for each individual in each activity. It is easy to deduce the number of units of time necessary for each of the individuals to produce a unit of any given good. Thus, the time which is necessary to an individual for the production of a unit of any given good is known. But in order to determine those activities in which individuals will be specialized, the wage rate per unit of time must be known. Lacking demand curves, Losch assumes arbitrary wages, which satisfy the equilibrium conditions in such a way that no worker would be excluded from interpersonal exchanges by reason of an excessive wage rate, and that the relationship between wages be consistent with the differences in productivity. Finally, by multiplying the time taken by each individual in the production of one unit of each good by these wage rates, Losch obtains the comparative costs of the goods
according to whether they are produced by one worker or another. Each individual will tend to specialize in that profession for which he possesses the greatest comparative advantage: that is, the profession in which he produces at the lowest relative cost. The logic of this choice of individual activities thus demonstrates that the subjects are specialized not only according to their aptitudes, but also according to earnings and length of employment, since these three variables jointly determine the highest wage per unit of time that the subject can obtain from full employment. The reasoning remains the same in the case of a limited monopoly where the worker is able to engage in several activities and to influence their incomes. Finally, the difference between the international and the interpersonal division of labor is only one of degree. While countries generally possess a capacity for production that allows them to export more than one product, no one individual is capable of satisfying the demand for most goods. In each case, the rate of exchange and the wage per unit of time fulfill exactly the same function. All this scheme implies that the decision of the worker depends solely on monetary considerations; but according to

Losch, the theory of comparative advantage can be extended to the calculation of the distribution of efforts and satisfactions.

The solution for the problem of the choice of individual activities may also be applied to the second cardinal question about the division of labor: that is, the determination of the personnel of a given activity. It becomes evident that a profession will not attract individuals who are comparatively better gifted, but will attract those who have in common the ability to maximize their money profits or their satisfaction, taking into account the wages paid and the length of employment. The marginal worker is thus not the least skilled worker, but the one whose hourly wage most nearly approaches his earnings in another occupation. The picture changes in those cases where fixed salaries or payments for a given performance are substituted for hourly wages. Furthermore the theory implies that individual productivity is not affected by the cooperation of workers. Losch relates productivity to the degree of utilization of the capacity of the firm, and the choice of activity then depends on the production possibilities of the worker as combined with other factors of production.

Proceeding to the examination of the third problem mentioned above, that of the locations of the individuals, Losch distinguishes between workers and entrepreneurs. At the level of the employed worker, only a very general principle is suggested: the worker locates at a point such that cost and disutility of moving to a new location cannot be compensated by the advantages to be gained at this new location. This is so, since money or real salaries are not sufficient criteria to measure utility and subjective satisfaction is not comparable over space. The location of the entrepreneur is similarly determined, but with the added condition that the location of the firm has a relatively greater importance to him.

The fourth problem is the determination of individuals that live in a given locality. In the wider sense, a locality gathers together all consumers and producers who achieve their highest utility there. In a narrower sense, a locality is a site which will go to the individual who is prepared to pay the most for it. This implies that, for given price relations, the individual in question finds his highest utility there, and that at the price paid for this site, he alone can achieve his highest utility there.

Examining the question of the industrial base of a locality, that is the fifth problem, Losch again finds two possible extensions of the concept of locality. In the wider sense, a given locality attracts those industries for which there is an adequate market for their products. Thus Losch leaves aside once more the principle of comparative costs as factors that explain regional and interregional specialization. In a narrower sense, a locality is occupied, by virtue of the above argument, by the firm which consents to pay the most for it.

Finally, the solution of the sixth problem, that of the choice of a location for a particular activity or industry, leads Losch back to the previous developments in the theory of location and economic regions.

In a self-regulating system, the spatial division of labor is determined by three kinds of local differences in prices: differences inside a single supply or market area; differences inside a single network of such areas; and differences between the different systems of networks and the networks of systems. Eventually each price reaches a level such that the conditions of locational equilibrium are satisfied.

After analyzing the nature of the division of labor and the theoretical conditions which lead spontaneously to equilibrium, Losch goes on to describe the mechanisms that assure a return to equilibrium when the system has been disturbed. Disturbances can be corrected either by transfers of products, or by the redistribution of the factors of production. Losch introduces, in the first case, his theory of "transfers", or the analysis of the flows of goods and services in the short run, given fixed locations. In the second case, he introduces his theory of "combinations", or the analysis of relocations of economic activities in the long run. But these two fundamental cases are studied in a more general framework, Losch having subordinated their analysis to the major distinction between a spontaneous system and a system of intervention. Finally, "transfers" and "combinations" appear both as regulating and automatic mechanisms, and as possible means of intervention.

In a spontaneous system, the flows of goods and services thus constitute the first mode of self-regulation (Selbstregulierung) or return to the equilibrium of the system. Any modification in the division of labor between two locations implies price fluctuations which play a definite role. This is demonstrated successively at different levels of observation: individuals, firms, regions and nations.

If a producer sees the demand for his services rise, he tends to modify his activities, increasing his prices and expenditures. These transformations of his own "balance of payments" bring about induced effects in the direction of prosperity with respect to his clients and suppliers and in the direction of depression with regard to his competitors. The same process is true for a firm. A variation in demand places it at the center of a series of price waves. Losch describes the rate and direction of the spread of these waves and the reasons for their dampening. A rise in demand, for example, produces a rise in prices whose spread is limited by the cost of transport and neighboring competition. The spread of the price rises is both direct, through the reduction of the market area and the enlargement of the supply area, and indirect, through the repercussions on the firm's competitors and customers. The direction of these price waves conforms to the series of transfers of purchasing power, which are themselves oriented toward the places where the prices have not yet been increased. As the price waves move away from their point of origin, their amplitude decreases; and with distance the influence of the original focus diminishes. The variation in purchasing power tends to be dispersed both between the different sellers and buyers of heterogeneous goods and among the buyers and sellers of each of the goods. Finally, to this leveling of the price waves in the successive periods of time is added the mutual cancellation when two waves of prices meet, since then the differences in purchasing power tend to compensate each other. Moreover, at the regional level, the price waves, although especially pronounced around the original focus, have a tendency to be reflected by the central places, or capitals over their own hinterland and transmitted directly to the central places of neighboring regions. This spatial transmission effect is strengthened by the banking system that undertakes money transfers between spatially separated economic capitals. Losch then shifts from the region to the nation, which is envisaged as a monetary space. Under a gold standard, effect of the credit pyramid is to amplify, uniformly within the jurisdiction of each bank, the effect of a local debt or credit on prices by induced variations in the monetary circulation. To the price waves is then added a displacement of the general level of prices; and the real quantities of goods exchanged exceed the necessary volume only by enough to compensate for the initial debt. This excess causes, in turn, a reflux of payments and exchanges and the general level of prices goes back to equilibrium level. There is a double transfer: a "preliminary" transfer, which is brought about by the displacement in the general level of prices and dampened by the reaction it causes; and a "necessary" transfer is tied to the price waves and becomes permanent once purchasing powers reach equilibrium. In the course of his analysis, Losch takes into consideration the differentiation of national currencies, and through the heterogeneity of monetary spaces finally reaches a theory of international trade. The levels of observation are not rigorously defined, but the fusion of interregional and international trade that seem responsible for it is neatly demonstrated. In effect, the disparity of the national monetary structures affects the characteristics of preliminary transfers by amplifying or dampening them, but this disparity does not substantially change the mechanism of necessary transfers. It modifies the general level of prices and the amplitude of price waves, but it cannot reduce the necessary transfers.

Losch's main innovation is the differentiation of the necessary from the preliminary transfers, which classic theory had blended. Here lies the originality of Losch's solutions to the main problems posed by the classicists. The temporary nature of the preliminary transfer guarantees that the international debt of a particular individual will not burden definitively his community. Nevertheless, the participation of the whole country belonging to the same monetary space in the transfer mechanism shapes the price fluctuations in the regions involved, independent of the fact that the preliminary transfer can be limited by the credits or deliberately controlled. The incidence of the transfer on those directly involved, moreover, can be regarded by the debtor as a gain in his role of consumer, due to the lowering of internal prices, and as a cost in his role of producer, due to the rise of external prices. This incidence varies as a function of the distance of the price wave from its origin, and is therefore proportional to the size of the nation. Losch compares a devaluation to a rise in the discount rate and this opinion leads him to distinguish between this devaluation made so as to facilitate transfers and one made with the intent to revive the economy. He finds that the choice of currency in which a debt is payable is important only when the rate of exchange is unstable. In the case of unilateral waves of purchasing power, the rise in price, lacking a compensating reaction, is definitive. Losch reconstructs an automatic equilibrium theory of the balance of payments, whose originality lies in the modifications he makes in its basic structure.

To monetary disparities can be added, moreover, customs barriers on the international level. But they only result in making the necessary price fluctuations even larger in order for the transfer to take place. However, the regions affected by a transfer do not necessarily coincide with the non-economic regions, particularly with political regions. Losch follows the configuration of the price waves beyond non-economic frontiers, over the whole globe.

If prolonged disequilibrium of necessary transfers persists, it can only be corrected by the migration of factors of production. This leads, in the long run, to new spatial combinations of the factors. In Losch's "combination problem", the locations become dependent variables and are regarded as the displacement of economic activities through which continual disequilibria, caused by changes in taste, technical progress, etc., are spontaneously corrected. Persistent deviations of prices in the countries where the purchasing power has changed are substituted for temporary price disturbances. The case of nominal and real wages is similar, resulting in a migration of labor until wage differences compensate for the disadvantages of change of place. Similarly, in the case of interest rate, capital movements accompany the migration of labor. But the demand of countries receiving immigration grows so that the migration of factors does not preclude the movement of products. The effect of the change on any one location on the locations of other activities depends on the fraction of their revenues that each of the countries derives from its emigrants and from other countries in general.

Having abandoned the classical assumption of the international immobility of factors of production, Losch examines the consequences of these relocations on trade and the international division of labor. There is nothing that guarantees an equalization of wages in space or time under these circumstances. On the contrary, migration tends to accentuate the relative disparities between countries, as it causes wage increases in countries with net immigration and declines in countries with net emigration. Moreover, the imperfection of mobility perpetuates the absolute differences among the wage levels in the same country. Finally, the participation in international competition of relatively less developed countries can be explained not by the principle of comparative costs, but by the kind of spatial protection that they find in some local factor of cost compression or in the cost of transportation. In this manner, Losch's theory of "combinations", as distinguished from the theory of "transfers", is opposed to the classical theory.

However, a spontaneous system does not guarantee an equilibrium that will necessarily produce the greatest possible social product or a given social product at the least cost. Losch finds two reasons for regulation of free equilibrium and he passes, then, from the study of auto-regulated mechanisms to the study of the external regulation of the system (Fremdregulierung). Depending on whether the intervention concerns the transfer of products or the combination of factors, he proceeds with the analysis of two types of regulation.

The regulation of the transfer of products avoids the problem of returning to the original equilibrium or facilitates the process of adaptation, or improves equilibrium position.

The first approach covers the extreme cases in which the process of adaptation is purely and simply suppressed. The prohibition of the movement of gold can prevent variations in the general price level; these variations are not necessary. In the limit, the price waves themselves can be dampened by the two contractant countries and goods transfer made impossible. In the second approach, a policy of raising tariffs can substitute factor immigration for the product movements that are associated with real transfers.

These two types of intervention, through monetary or tariff regulation, are found in those cases where the adaptation process need only be facilitated. Monetary techniques appear to be the most important ones and their means are numerous depending on whether the State acts within the framework of existing institutions or whether it modifies the institutions. Losch proceeds to list the advantages and disadvantages of the gold standard and of the paper currency standard, with regard to their primary effects on the changes in price levels and their secondary consequences, taking into account the importance of the percentage of gold backing of the currency and of the permanence of the changes. He then analyzes the different means by which central banks can influence price levels and exchange rates. Under the gold standard, to the extent that the size of the gold reserves allow it, the central banks can weaken or amplify the price variations and the exchange rates and, therefore, the corresponding preliminary real transfers, by applying sufficiently flexible discount policies. Depending on whether or not these policies weaken or reinforce the efficacy of gold in the transfer mechanism, they in turn determine whether larger or smaller quantities of the metal are regarded. The effects of the discount policy vary according to the importance of the legal gold reserves and the unilateral or coordinated character of its application. The gold standard can accomodate itself to a variable metallic coverage and a constant exchange rate, or the contrary. In the first case, the price level can be stabilized by the debtor country as long as its gold reserve is not exhausted; in the second case the movements of gold and of the exchange rate occur simultaneously. The paper standard currency offers the same possibilities for manipulation as the gold standard, especially when associated with a foreign exchange reserve. However, in the case of great disturbances, these diverse techniques have a limit to their effectiveness after which the monetary institutions themselves must be changed. The adjustment can then take place either by the variation of the rate of exchange or, on the contrary, by exchange control. In addition to the monetary policies, tariff intervention can ease the transfer process. Adjustment of the tariff permits the substitution of the export or import surplus for the variations in prices and exchange rates. An inverse adjustment can also amplify the price movement needed for the transfer of goods; even, it can substitute migration of capital for product transfer.

The regulation of the free market may have, as its objective, the improvement of the stabilizing effects of the transfer. Given this approach, it is convenient to specify the desired objectives. Losch distinguishes three: higher business profits, full employment and autarchy. Then new methods of regulation become available through direct control of prices, quantities, and values of imports and exports. The essential process of a final transfer, following a preliminary transfer, is the same whether the mechanism of adjustment is spontaneous or controlled.

However, regulation may be used, not to modify the real and monetary flows, but rather to rearrange the combinations of factors of production used. As in the case of the regulation of transfers, the regulation of combinations may be directed to the accomplishment of any one of three objectives: the expediting and easing of the process of adaptation; the regulation of the return to equilibrium; or the improvement of the equilibrium position.

The process of adaptation can be prevented by the preservation of old spatial relationships and the prevention of new combinations. The complete immobilization of the existing locations by appropriate force, payment, or education allows an economic landscape to be fixed although it is, by its very nature, unstable. The economic justification of these policies is different depending on whether or not they result in an increase in welfare. The impediments which are deliberately raised to prevent new spatial relations appear to be more limited in their effects. The reinforcement of commercial barriers has as its function, on the international level, the prevention of a new internal distribution of the factors of production when the international division of labor would lead to an undesirable specialization. Similarly, the prohibition of the migration of factors can oppose international readjustments, yet it risks the substitution of the movement of products; or, if it is not absolute, the use of new factors complementary to those that escape the prohibition.

In the case where regulation need only facilitate the process of adjustment of factor combinations, the difference existing between Losch's model of optimal location and the real world must be reconciled. This difference is due to the lack of the conformity of actual conditions to certain fundamental theoretical assumptions: rationality, mobility, and competition. However, appropriate measures may be applied to diffuse information so as to improve the rationality of the individual decisions and increase the mobility of decision makers, thereby making the actual system more freely competitive.

Finally, regulation may have a more creative aspect in encouraging more productive factor combinations. Creative regulation can be applied to promote the expansion of a people, a space, or a given industry. In order to analyze the diverse characteristics of the growth in prosperity of a nation, Losch discusses three possible levels of spatial expansion: colonization; the creation of large economic areas; and the establishment of economic unions. While the vigor of a given nation demands that the management of its space be subject to regulation policies, the fact that the expansion of a given space is given a privileged position goes back to the consideration of the spread of the population as the purpose of the regulation. The types of actions by which the actual location will be brought closer to the ideal depends on the level of development and on whether the attempt is to optimize the spatial organization of an old country or of a young one. Finally, if only the expansion of a particular branch of the economy is desired, all the possible regulatory means are once again available, from simple education to force. Generally, the principle of regulation is based on the idea that if several means can be useful in the realization of a single objective, several objectives can eventually be obtained by using only one means.

CONCLUSION

Losch's work shows decisive progress when compared to the writings of his predecessors, notably Palander and Ohlin. Losch's attempt represents the first modern and systematic integration of various earlier theories into a unified analytic structure. It outlines not a unique scheme of explanation, but a set of coordinated models each appropriate to the interpretation of a spatial problem at one or another level of observation. The idea of equilibrium, which constitutes the unifying link between the specific models, serves also as a link with general economic theory [236]. Thus, even though his theory of economic regions rests on a set of restrictive assumptions and is a special case of general theory of interdependence of locations, Losch satisfies the goal of spatial theory. This goal, which was particularly clear after Predohl, was to put an end to the historical division between spatial theory and general economic analysis.

Finally, the high level of abstraction on which he places his theoretical developments should not overshadow his more concrete contributions. The last part of his book is entirely devoted to empirical illustrations (Beispiele) using precise examples to show the application of the various abstract models. In general, the majority of Losch's writings are based on important real world observations. By this constant effort to tie the most abstract analysis to a concrete interpretation of reality, Losch makes possible significant methodological progress in spatial analysis.

Nevertheless, the objective of the generalization that he attempts is, to a certain extent, incompatible with his particular method. The need to be systematic leads him to an extreme rigor of thought and expression, but the part assigned to the analysis of regions and to the division of labor introduces an imbalance in his whole structure. In a sense, it may have been premature to integrate the more original contributions of Losch into a critical and exhaustive exposition of established theory. To the extent that Losch considers his last book as the presentation of his preliminary results, he himself attests to the provisional character of his conclusions.

The dramatic death of the author on May 30, 1945, at the age of 39, gives his work its unfinished character. In his last days, Losch was engaged in the development of the monetary aspects of his theory (7). However, despite leaving his work unfinished, Losch occupies no lesser place in the history of spatial theory than he would have deserved had he completed the development of his entire theory (8).

---

(1) Losch (A.): "Wirtschaftsgebiete als Grundlage des internationalen Handels", unpublished lectures given at the University of Bonn, 1936.
(2) Cf.Mathematical Appendix IV,1.
(3) Cf.Mathematical Appendix IV, 2.
(4) Cf.Mathematical Appendix IV, 3.
(5) Cf.Mathematical Appendix IV, 4.
(6) Cf.Mathematical Appendix III, 4.
(7) A book project found in his papers has been the subject of a posthumus article [162].
(8) A bibliography of Losch's works was published in 1971 [455]. Essays in honour of Losch were published in 1978 [637].

Chapter 9. From Losch to the Nineteen Fifties

In the history of spatial analysis, Losch's death marks an important date. Its importance is not only derived from the fact that his premature death deprived this field of its most eminent theoretician. The importance of this date stems from the dimensions of Losch's work. The monumental architecture of his construction gives it a definitive character. While based on a particular system resulting from the intellectual bias of the author, the synthesis is already sufficiently general and comprehensive to integrate the majority of the previous theories. From this point of view, his work appears to be less a consumation of Losch's thought than the result of a century of analysis. But, even more, to the extent that Losch's contribution postulates some new developments, his work foreshadows the majority of later studies. He appears, thus, as the landmark in modern spatial research; and the definitive character of his work is confirmed by the fact that it marks a new level of analysis.

However during the years after the second world war the German school of thought lost its hegemony and was superseded by Anglo-Saxon thought, more specifically American thought.

The first distinguishable period is that going from Losch to the nineteen fifties. An overall appraisal of spatial economic theories can still be made up to about 1955. However from the late nineteen fifties spatial economics embarked on a course of very rapid and diversified development, which is still in progress. This contemporary period lies outside the realm of an historical study. A certain lapse of time will be necessary before the really important works stand out, thanks to the lasting influence they have exerted, and before rediscovering the works of possible pioneers who were ignored or misunderstood by their contemporaries.

From 1940 to 1950 nearly all abstract research was applied to the study of partial equilibria. 1950 to 1955 was a period of relative diversification in the scope and methods of spatial economics. Most authors focused their attention on general spatial interdependence by either employing a general equilibrium model or by analysing its component parts such as, the entire transport system or the relationships between economic regions.

Section 1. PARTIAL SPATIAL EQUILIBRIA

Partial equilibria analysis developed basically between the two successive editions of "Die raumliche Ordnung der Wirtschaft", i.e. from 1940 to 1944. Although the different analyses were contemporary with Losch's main work they were not influenced by it, quite the contrary. Indeed they are the continuation of much earlier thought going back to Launhardt, Fetter and Hotelling.

The numerous authors that have made contributions in the field of partial equilibrium have made progress in two general directions. On the one hand, the fundamental assumptions of the previous models have been the object of a meticulous examination and a critical appraisal to find the limits of their validity. On the other hand, as a consequence, the majority of the conclusions previously attained have been generalized, as fewer restrictions have been introduced or additional hypotheses suggested.

From this point of view, the study of duopolistic market structures seems to have attracted the attention of the majority of the researchers. Hotelling's theory is examined anew, first of all, by Austin Robinson [124] and Arthur Smithies [127]. These authors observe that Hotelling's law which states that duopolists tend to concentrate their locations at the center of the market is not valid for assumptions other than those made by Hotelling: that is, an inelastic demand, zero cost, and a constant rate of transportation per unit (1). Following the extension of this model that Lerner and Singer had already developed by considering an inelastic demand only for a specified range of prices from zero to a given upper bound, Robinson and Smithies generalize the model by introducing an infinite elasticity of demand and by analyzing the effect of variable costs. In a more systematic way, Smithies also examines the effect of variations in the cost of transport. In all these cases, a tendency contrary to that of concentration appears: to the centripetal force, described by Hotelling, is opposed a centrifugal force and the law of concentration applies only to a particular case.

The Norwegian economists, C.D.Hyson and W.P.Hyson, undertake a critical review of the models that followed the tradition established by Fetter of having the location of the duopolist given [175]. In these models the shape and size of the market of each duopolist are the dependent variables of the analysis. "Fetter's law of market areas" states that the boundary line between two competing market areas becomes a hyperbola which is the locus of points where the difference between the cost of transport from the location of each of the two sellers is equal to the difference between their prices. This law also applies only in the particular case where the rate of transport per unit of distance is assumed identical between all points in the area under consideration. The introduction by Hyson and Hyson of unequal transport rates for each seller leads to an equation for the boundary line that corresponds to that of the "hypercircles", the family of curves representing the locus of points that have a constant ratio of distance from two fixed circles. Depending on the ratio of prices and the ratio of transport costs of the two sellers, the curve can become not only a hyperbola, but also a circle, or a perpendicular bisector of the line joining the locations of the sellers. But following this analysis, Hyson and Hyson could only concur with the conclusions of Schneider and Palander who had already generalized Fetter's theory in their systematic analysis of "isostantes".

Of course these elaborations of the theories of location strategies and market areas are not able to cover completely the multiplicity of possible situations corresponding to an infinity of complex combinations of assumptions. Thus, Smithies examines a system of multiple "basing points" and discusses the policies of identical duopolists located symmetrically at two base points in a linear and uniform market [132]. The assumptions of identity, uniformity, and symmetry are then relaxed and the problem of the "entry" of new competitors, without a base point, is examined. Numerous concrete cases of basing point systems existed in the United States, giving this analysis a greater flavor of reality [152][153][157]. Similarly, the assumption of the continuity of the market, employed since Launhardt in all the models that follow the tradition of Hotelling or Fetter, is questioned by Ackley [130] (2). The buyers who are spaced along the market line are divided into groups whose particular demand curves differ. Ackley studies the different possible price strategies displayed by duopolists whose locations are given and fixed.

In contrast to duopoly, other market structures have received relatively little attention.

Smithies is the only one to have approached the case of monopoly [126]. In reality, in spatial economics the assumption of a pure monopoly would only cover very particular situations, since it implicitly presupposes the non-contiguity of the market areas of competing sellers. That is, the monopolist's market area is completely isolated from all external competition either because of distance or because of an institutional boundary. Nevertheless, Smithies shows how the shape of the demand curve determines the way the monopolist fixes his price. The maximization of his profit leads him to practice f.o.b. or uniform delivered pricing or to absorb a portion of the cost of transport according to the reactions of demand to price. The limitations in the application of this model are fairly strict, since the distance factor constitutes the only independent variable in a model where the location on the monopolist at the extreme of a finite linear market is given and fixed, the demand decreases as the price increases, and the costs of transport are uniform per unit of distance. Finally, no discrimination in prices appears, since it is assumed implicitly that the demand is elastic.

Similarly, Enke is led to a reconstruction of the theory of imperfect competition by integrating the spatial factor into the analysis of competitive markets [131]. Enke starts from a criticism of Marshall's conclusions that the addition of the cost of transport to the competitive price, resulting from the interplay of supply and demand, determines a whole series of local prices. The transport portion of these local prices depends on the price of the product delivered to the buyer. Enke, after showing how this approach covers only the particular case where the sellers are concentrated and the buyers are dispersed, generalizes the analysis by examining systematically the formation of prices according to the different possible assumptions about the concentration or dispersion of the respective locations of the contracting parties. In this manner, he shows, contrary to the theories of imperfect competition, that the spatial factor cannot be simply reduced to a number of frictions that prevent competition from being perfect. Moreover, in his combinatorial analysis of location, he foreshadows Miksch's theory of spatial equilibrium.

Monopolistic competition may be distinguished from imperfect competition; and in 1951 it is Chamberlin who approaches the problem of the relationship between the spatial factor and the impurity of competition [181]. More explicitly than in the first formulation of the theory of monopolistic competition [75], the inelasticity of demand is related, in the more recent expositions, to a number of assumptions about the location of the population. Similarly, the size and number of firms depends on the cost of transport and thus on the area of the markets. Also, distance spatially differentiates even homogeneous products and advertising modifies the buyer's sensitivity to distance. Nevertheless, in as much as space still remains a simple element within the category of product differentiation, its integration within the analysis of monopolistic competition remains only partial. Due to this fact, the differentiation of price is not postulated in the model. In a more general way, Chamberlin's analysis, contrary to that of Hoover and more specifically that of Losch, does not integrate completely the spatial factor into the theory of monopolistic competition. Moreover, even though in Chamberlin's exposition the firm constitutes the fundamental unit of the economic structure, his analysis is at the level of the mutual interdependence of the units of the system.

At the level of partial equilibrium, aside from an interesting application by Lewis of models of the Hotelling type to the retail markets [141], only the labor market has drawn the attention of theoreticians. Goldner, taking up a previous study by Reynolds, systematically studies the implications of the relative immobility of labor and the nature of the boundaries of the labor market [144] [238] from the point of view of supply and demand.

Section 2. GENERAL SPATIAL EQUILIBRIUM

Contrary to studies of partial spatial equilibria, the analysis of general spatial interdependence with all the corollaries inherent in its fundamental theorems follows on directly from Losch work. The reason is that Losch had formulated, without developing it, a system of equations by which the conditions of general spatial equilibrium can be expressed analytically.

The elaboration of a theory of general spatial equilibrium is the result of numerous and diverse contributions. Nevertheless, the apparent disorder that has resulted from the variety of approaches used by various researchers must not overshadow the basic unity of thought underlying the continuous development in this field.

The work of Isard engendered a long series of successive contributions. Isard's preoccupation with basing his theoretical deductions on an adequate set of empirical observations is reflected by the extent of his empirical research. This research resulted in frequent changes in the level of abstraction of his studies, changes which are subordinated to the search for an abstract analysis of the general interdependence of the spatial system. His theoretical contributions culminated in 1956 in a major reformulation of the whole theoretical framework [257].

In 1949, in a first synthesis, Isard shows how spatial analyses tend to include the modern theory of general equilibrium as a special case. Non-spatial general equilibrium is merely spatial general equilibrium where the costs of transportation are assumed to be zero and the factors and products are perfectly mobile [160]. Losch's general equations had presupposed a formulation which would be able to express all the relationships of substitution existing in the system among the costs of transportation, inputs, and outputs. In order for the equations to be developed further, the principle of substitution, which Isard borrows from Predohl, seems the most appropriate instrument of analysis for a complete description of Losch's spatial configurations. From this arises the need for a rigorous definition of the basic concepts which would allow such a description to be made and which would fuse spatial analysis and modern orthodox economic theory. Of these concepts, the most important one contributed by Isard is that of "distance inputs" or "transport-inputs": the physical cost per unit of distance of moving a unit of weight of a good. Transport-inputs allow the attachment of some sort of spatial coefficient to all inputs and outputs. More precisely, the relationships of interdependence of an economy can be described with the help of transport functions whose corresponding inputs are the "transport-inputs". The transport functions must also be combined with the other functions of production and consumption that express the relationships of the system envisaged [188].

Isard is then able to formulate mathematically the set of equations describing the conditions for an optimal spatial equilibrium of an economy under certain restrictive conditions. These conditions are the continuity of the means of transport and of the market and supply areas [207]. He is thus led to integrate and at times to generalize within the same framework a number of earlier partial theories of the Weberian type, of market areas, or of agricultural location. In particular, the former dichotomy between agricultural and industrial location, which had persisted in spatial analysis up to Losch, is finally overcome. Thus, Isard specifies explicitly the conditions of general equilibrium for the set of locations of a system.

The primary importance of this work of Isard results from the fact that it completes Losch's model, by deducing all its theoretical consequences and following its development and application to the most advanced fields of contemporary research. Isard appears to be the chief trustee of all the traditions of spatial economics and of this intellectual heritage, thereby gaining a central position in modern thought.

At the same time, Miksch shows how in a state of equilibrium, locations order themselves according to a limited number of typical models, how these elementary models tend to articulate themselves with one another, and finally how the optimal static equilibrium position is only a tendency when dynamic considerations are introduced [178][193]. The study of general equilibrium follows in three main directions. First of all, Miksch puts greater emphasis on the description of the forms of equilibrium than on the conditions of this equilibrium. Secondly, he tries to show how equilibrium is reached by the combination and interrelation of typical elementary models. Finally, the position of optimal equilibrium described by a static model appears only as a point of reference in a limiting case within a dynamic model where only relative positions of equilibrium exist. This is because the rigidity of the previous locations delays the necessary transfer of investments.

The analysis of general spatial interdependence, as discussed by Isard and Miksch, has additional interest because of the new directions it opens up. Advances in these directions had previously been dependent on a sufficient knowledge of the relationships existing among the set of locations of the system.

First of all, such an analysis permits the location of the transport network to be accounted for. Generally, in previous models, the transport networks were given in the analysis. More often, the authors assumed that transportation between any two points following any line was always possible. This ubiquity of communication lines was only questioned by Palander [87] and Stackelberg [106] in their study of "the law of refraction", in which the number of points and lines of transport depended on the means of transport envisaged, while the location of these points and lines depended on the need for transport. Only Losch's theory of the system of networks had permitted the simultaneous determination of the location of economic activities and that of communication lines and traffic intensity in each. But Losch's theory of regions, limited as it is to a system of hexagons or squares, still makes impossible a model of communications of the level of generality reached when the communications model becomes subordinated to a real theory of interdependence among a set of locations.

In the models of Hitchcock, Beckmann, Dantzig, Dewey, Koopmans and Reiter, the spatial distribution of economic activities is given and optimal and the transport networks are deduced [119] [161] [168] [182] [191] [200] [201] [202] [213]. The optimal transport model which results then incorporates the old "law of refraction" as a particular case.

However, due to its generality, the optimal model has only a limited application in those cases which satisfy the assumption of continuous spatial distribution of the points served.

Beckmann, McGuire, and Winsten construct models based on more realistic assumptions, allowing the problem of congestion to be treated [250]. Such an analysis can then take into account the variability of transport costs and the indivisibility of the means of transportation.

To the extent to which Isard's general theory is an application of a Walrasian model to the analysis of a system in which distance is a variable, it could not fail to lead to the adaptation of input-output models to regional data. The relationship that ties Walras' coefficients of production and Leontief's technological coefficients leads to the possibility of a matrix description of interregional flows. In his first approach to this problem, Isard envisages the disaggregation of a national economy into categories of activities, each of them being undertaken in each region [189][190]. The construction of an input-output table leads to technical coefficients which are differentiated according to regions. But, in the path thus opened by the initial formulation of the author, two other kinds of models result.

First of all, it is possible to pass from this interregional model to an intranational model. Instead of starting from the interregional structure of different types of national activities, Leontief and Isard formulate this second model by using the national coefficients. They base the distinction between those activities which are regional or national in character on the distances that the products of each type of activity are transported [209]. In other words, the interregional model rests upon an aggregation of local quantities. It is therefore of use in determining the national implications of the structure of activities in each region. The intranational model rests, on the contrary, on a disaggregation of national quantities. It is of use in determining regional implications of the interindustry structure of the national economy. The intranational model is thus complementary to the interregional model. Finally, there is a third, less ambitious method which has also been used by Isard, Kuenne and Freutel. It consists in applying the interindustry national model to any one region, to a city or to a group of regions.

No matter which particular methodology is used, all matrix analysis leads to a description of the economic system in terms of interregional flows. This technique, directly derived from Isard's general spatial interdependence model, is the result of one line of development of the theory of economic regions. But with respect to this theory, the development from Losch to Isard must include Vining and Neff. Shortly after Losch completed his deductive exposition of regional structures, Vining presented a descriptive exposition of these structures, which constitutes an important step in the elaboration of the modern analysis of interregional flows [143][148][149][151][163][166][226].

Preoccupied with the explanation of the regional disparities in the effects of a national cyclical fluctuation, Vining shows that the measure of the rate of variation of the characteristic size of the national economy is identical with the estimation of the parameter of a frequency distribution of the rates of variation of the same size measures in the various regions of this economy. In order to account for the differences that appear in the rate of variation and the turning points of different regional series, Vining studies the relationship existing between the location of dominant industries and the degree of regional specialization of various activities. He goes on to consider a series of interdependent regional economies, each of which possesses a definite commercial circuit. The sensitivity of a given region to changes exogeneous to it therefore depends on that circuit, through which it receives and sends impulses. The region then appears to be an organic complex. It becomes, similar to Losch's region, a significant spatial aggregate: that is, the place of the reception and emission of a whole set of input and output flows. It possesses a balance and its sensitivity depends on the interregional trade multiplier. Vining derives the formula for this multiplier from that of the international trade multiplier.

Through Vining, the theory of economic regions rejoins modern theories, derived from the work of ecologists, in which the region is characterized by the presence of a dominant focal point and a satellite area. The ties subordinating the dependent locations tend to weaken as the distance from the nodal point grows. Beyond a certain point, greater strength will be observed in the likages to other central points. Finally, the hierarchy of regions around a dominant metropolis can be constructed as in the theories of Christaller and Losch [208][217].

It is not by chance that the theory of the general interdependence of locations and the theory of regions reach a high level of development simultaneously. They strive for a common goal: the analysis of networks. Their differences, which appear to be minimal, are methodological, and can be explained by the origins of the theories. The theory of general interdependence expresses and develops the system of equations through which Losch had formulated the general conditions of the optimal equilibrium of a spatial system. It is based on a combinatorial analysis which shows how all the sites are located with respect to one another and how they are ordered according to well-defined models. The theory of regions, in the strictest sense, develops from the analysis of Losch's system of the network of areas. It starts from an aggregative conception, determining simultaneously the sets of locations and their structures. In both cases, the analysis is at the level of economic regions, in the largest sense, that is, of the network of locations linked to common centers, themselves interdependent and arranged hierarchically (3).

The corollaries that can be derived are related to the theory of interregional and international trade.

Ohlin's failure left the theory of interregional trade at an impasse. He was only able to envisage interregional trade as a palliative for the lack of the mobility of factors, when he attempted to bridge the gap between the neoclassical theory of prices and the theory of interregional trade. Losch, on the other hand, starting from this theory of systems of market areas and networks of systems, was able to base the theory of exchange on an analysis of the spatial division of labor. All the efforts of modern analysis point out the close relationship between the theory of location and the theory of interregional trade [264]. This reemphasizes the basic unity between the interdependence of location and the flows of interregional trade. It explains why the theory of interregional trade could not be developed much further than the point to which Losch had brought it, as long as a model of spatial interdependence was not available on which to base it. Beyond the methodological differences that distinguish them, the works of Miksch, Beckmann, and Isard show how spatial interdependence makes locations dependent on the prices that result from competition among users for raw materials and labor from several potential sources

[193][213][230][231]. There is complete correspondence in the model between spatial prices and material sources used. The distances are expressed in terms of local differences in prices and the location of economic activities in terms of flows.

The analysis in terms of flows generally tends to dominate contemporary spatial economics. No doubt because of its structure, the theory of spatial interdependence incorporated the theory of regions at the end of their historical convergence. But, because of this, the spatial problem is progressively confused with the problem of the allocation of resources and the maximization of employment. Enke solves the problem of equilibrium between spatially separated regions by a system of linear equations [184]. Samuelson treats it as a maximization problem and applies the method of linear programming to its solution [211]. His model, in which the total transport costs are unknown, appears to be more general and incorporates the problem of minimizing total transport costs developed earlier by Hitchcock and Koopmans, where unit costs are assumed given [119][161]. Similarly, Beckmann and Marschak use activity analysis to solve the problem of determining the volume and distribution of the production of an integrated firm for each of its branches whose locations are given [235]. No one doubts that these methods are particularly appropriate to a spatial analysis formulated in terms of flows. The application of these methods does appear to place location analysis in the mainstream of modern mathematical economics.

Finally, almost at the same time, Hoover and Meyer-Lindemann both ingeniously reconstruct a comprehensive presentation of location problems. While using their own particular methodologies they converge towards a definition of economic policy which is likely to guarantee maximum social efficiency to the system of locations.

Hoover attempts a comprehensive presentation of the spatial problem [154]. Adopting a level of abstraction inferior to that of the authors who inspired him (Weber, Palander and particularly Losch) he develops a scheme fairly similar to that of Dean, although more comprehensive [102]. Hoover starts from the problem of individual location, subordinating the study of the explanatory factors of local differences in cost of transport and production to the analysis of the total cost of supply, transformation, and distribution. Thus, still keeping a Weberian point of view, he is able to develop a more satisfactory technical approach. He treats systematically the relationships between transport costs, other costs, and the location of individual units and then goes on to study the interrelationships of all locations. His most original contribution is in the introduction of time into his analysis. The equilibrium of location appears only as an evolutionary tendency needing perpetual adaptations and constant readjustments. The specific problems of international trade are then treated as a result of the influence of boundaries, which emphasizes the fundamental analogy between domestic and foreign trade. Finally, Hoover goes beyond the point of view of private initiative and attaches primary importance to the state-determined framework within which private initiative must work. He is led to take stock of the objectives and means available for a rational policy of the planning of locations. In the end, he brings up the problem of a policy of international integration, although he does not deal with it. On this point, only Giersch has attempted a study of the spatial implications of economic integration and underlined the discriminatory influence of space [159].

Compared to Hoover's work, Meyer-Lindemann's seems more restricted in scope, although not in depth [192]. He limits himself to industrial location. Having elaborated a typology of specialized theories, he analyzes the use that could be made of them for economic policy. In his critical presentation of the theories of industrial location, he starts from the establishment of the firm at the point of minimum transport cost, criticizing the Weberian version of this problem. From the study of the determination of individual location, Meyer-Lindemann goes on to consider its effects, relating the theory of location to the theory of prices. To these abstract analyses, he juxtaposes the disparate contributions of the historical schools of Roscher, Schaeffle, and Ritschl, and of philosophers like Bulow. Finally, Meyer-Lindemann delineates the relationships of the firm and the system. He tends toward an analysis of the rational economic laws of the occupation of space. The technical and sociological factors are similarly approached, notably from the point of view of the consequences of industrial migrations.

In a general way, the German school is oriented towards a theory of planned spatial organization, conceived less from the point of view of individual motivations and the realization of a purely economic optimum than as a function of the collective advantage and the achievement of a general optimum.

Having started with a very different purpose, Predohl had made a similar study. After having indicated, without developing the argument, the possibility of enunciating a general theory of location in terms of substitution, Predohl applies this theory to the analysis of the international economy [164][177][195]. Nevertheless, he proceeds not to determine the points of substitution of groups of productive factors reacting to a displacement of production, but rather to outline the phases of the secular evolution of the world economic space. Substitutions are thus looked at totally and from a temporal point of view. Thus Predohl departs from the abstract theory of the general equilibrium of locations to integrate the spatial factor in the evolutionary model of Sombart. He uses the latter's historical categories to show how the specific models of the ordering of world space correspond to the three typical phases of the development of capitalism. Infant capitalism oriented the world economy around the European industrial nucleus and organized it in the form of a "uni-concentric" model (Uni-Konzentrische Weltwirtschaft). The expanding phase of capitalism, creating in North America a second industrial nucleus, oriented itself on a "bi-concentric" model (Bi-Konzentrische Weltwirtschaft). Finally, in a third phase called "intensification" in opposition to Sombart's concept of declining capitalism, Predohl sees the birth of a new center of industrialization in the U.S.S.R. This leads to a "tri-concentric" model (Tri-Konzentrische Weltwirtschaft). At the end of his work, Predohl poses the problem of the relationship between economic regions and political regions, especially from the point of view of full employment and growth policies [196]. But by accenting the tensions that result from the organic needs of economic regions and from considerations proper to political regions, Predohl enlarges the search for criteria for rational planning by underlining the necessity for the coordination and
harmony of spatial structure for the whole world.

(1) Hotelling's and Smithies'theories are summarized by Greenhut (M.L.) [205].
(2) Cf. the controversial Fetter-Hinrichs, Temporary National Economic Committee Hearings, March 7, 1939, Part 5, p.1917 ff.
(3) Cf. the synthesizing study by Popescu (O.) [245].

Chapter 10. Since the Nineteen Fifties

INTRODUCTION . THE SCIENTIFIC ENVIRONMENT DURING THE SECOND HALF OF THE 20th CENTURY

The nineteen fifties mark a turning point in the history of spatial economic theories. Since then spatial analysis has enjoyed rapid development in new conditions.

There were as many specialists alive during the period 1950-1980 as the total number of authors from Cantillon to Losch. Nearly all scientific disciplines have experienced a similar evolution during the second half of the 20th century. More and more works and new specialized journals have been published. Yet at the same time there has been a constant increase in the number of articles, devoted to spatial analysis, published in non-specialized economic, operations research and geographic journals. The findings have been gathered together and synthesized in a vast number of textbooks, in particular [297] [354] [355] [379] [382] [416] [417] [424] [428] [488] [570] [682]. Specialized research centers have been set up in the five continents. Scientists have grouped together in regional associations and symposiums have been organized regularly. Finally spatial analysis has become a regular university subject with the result that several textbooks have been published.

In this way a world-wide scientific community has emerged, dependent for its structure on the location of the research centers and on the circulation of their works.

Under the impetus given by Isard and his school, American thought has exerted great influence all over the world. It has been particularly dominant in Anglo-Saxon countries, the Federal Republic of Germany, Japan, Canada and in India. In French speaking countries, spatial economic theory, after having been ignored for a very long time, was introduced in 1955 by Ponsard [244], then by Moran [353] and by Boudeville who has limited his analyses to the polarisation of space [469]. However, certain authors still prefer the ideological controversies inherited from the 19th century, as Scott remarks concerning the literature on land rent [604]. Nevertheless, modern scientific analysis has spread widely both in France and Belgium. The Dutch school has also made an important contribution. As to the other Western European countries, spatial analysis has gained ground there to different degrees. Finally, research in the U.S.S.R., where doctrinal controversies over marxist-leninist accountancy prevailed for a long time [287], has been mainly devoted to regional planning [373] [408]. The situation in the Eastern European countries is rather more complex. Their preoccupation with the voluntary organization of space is coupled with an interest in broader topics, marked by Western influence.

Such a diverse scientific environment makes it very difficult to come to grips with all the currents of thought which have enhanced spatial economic theory. As research has become more diversified and specialized so the main axes have overlapped and the abundance of publications has somewhat clouded the main themes. Yet, this tangle can be unravelled by singling out two currents. Firstly, those works which are the continuation of the historical legacy of spatial economic theories. Secondly, those works which tackle relatively new subjects and which are, therefore, only indirectly influenced by history.

Some theoretical models have been used as the analytical framework for discussion and research into the problems they deal with. They have been recognized as authoritative models of reference and as such they now constitute the true paradigms.

New directions have emerged from other currents of research which, on the contrary, cover new or relatively unexplored areas of spatial analysis. It is true that their roots are to be found in long established concepts, however their development is relatively independent of the paradigms inherited from the past.

Section 1 . THE PARADIGMS

Four great theoretical models can be described as classical. Thunen's concentric rings, which, although they were in fact discovered before his work, still remain attached to Thunen's name and have proved to be remarkably durable. Weber's problem which was restrictive when initially formulated but has been progressively generalized. The solution to Hotelling's model of the concentration of locations has earned the status of a law. The different solutions resulting from the introduction of new assumptions are still presented in reference to that law. Finally, the central place theory of Christaller and Losch has served as the archetype for a comprehensive analysis of economic regions, their components and interrelations.

Each of those paradigms presents not only the structure of a given state of knowledge, but also contains its own problems and engenders new research. Thus each has its own internal dynamics which carries its historic evolution forward.

1. The perennity of Thunen's concentric rings

The "Isolated State" model has never ceased to exert a certain fascination on spatial analysts, be they geographers, like Grotewold [288] or Katzman [533] or economists like Clark [367] or Dickinson [403]. This continued attraction explains the permanence of Thunen's model which continues to serve as a reference in several works on agricultural location as well as offering a theoretical model to urban location analysis.

In spatial agricultural economics, when the model is applied in a growth context, it can be shown that growth modifies the order and content of concentric rings and causes several town-centers to appear [241]. Thunen's statically built model calls for a dynamic formulation [402].

Yet Thunen's model enjoys only limited descriptive weight. Jones, Mc Guire and Witte, in a deeper study of its structure, have specified the transportation costs and their impact on land rent [645]. They look into the validity of the empirical explanation given by Thunen for rings with different intensities in the production of grains and the relevance of his laws of agricultural location. They show that Thunen's laws are limited in scope because they do not take demand conditions into account and they do not meet the conditions of boundedness.

However when one acknowledges that fertility is not a natural phenomenon then new life is given to the analysis of location and land rent in a concentrically structured space. One piece of land is better than another and yields a higher rent because a combination of factors of production, implying a lower unit cost, is used there. The hierarchy of rents and fertilities depends on the productive combinations and the distribution of wealth. This study, which was originally developed in the framework of non-spatial equilibrium models, inspired by Sraffa and using a single agricultural product, has been transformed by spatial analysis. Scott goes into the spatial distribution of production techniques as a function of the rent they enable one to obtain [605] [693]. Then Huriot takes Thunen's model further with the help of a linear and plurisectorial production model [729]. He gives several fundamental findings: the location of marginal pieces of land depends on the distribution of incomes; the same product can be produced in several non-contiguous rings; in general land rent does not drop regularly when the distance from the center increases; the locations of production techniques depend on the relative intensities of the use of factors characterizing them, on the price of those factors and the price of the goods produced. In agricultural economics, the different production techniques can simultaneously produce each good corresponding to the several qualities of land available. However this type of approach is no longer restricted to the agricultural sector. In urban economics, the choice of a production technique no longer depends on the quality of the land, but remains determined by the distance from the center, by prices and by incomes.

From a different standpoint, certain specialists in urban economics have taken up Thunen's model to provide spatial urban analysis with its missing theoretical framework. Seminal works in this field have been published by three authors, Wingo [306] [307] [323], Muth [301] [410] and above all Alonso [324]. As in the Thunen "Isolated State", the town is a plane surface endowed with a privileged point, its center, from which urban locations organize themselves in concentric rings. The allocation of urban land to different types of use, consumption, residential and production, depend on the land rent. At each point in urban space it is the maximization of land rent which determines the nature of any location. Land rent falls with distance from the center such that the latter plays a similar role to that which Thunen had assigned to the distance from a market town.

Moreover, the dualistic nature of the respective location and rent configurations in Thunen's theory has led analysts to reformulate the classical theory of location with the help of programming models [395]. In particular, Herbert and Stevens have shown that the formulation of the Alonso's theory and that of the allocation of urban land program with which it can be associated, are mathematically equivalent [295]. So the theory of optimal control can then be used to dynamize Alonso's theory as well as that of Herbert and Stevens [591].

Finally, in order to find out whether or not the size of the market town is optimal we have to analyse the relationship between traffic congestion and the allocation of the land to transport and crops. Kanemoto shows that the market town is larger than the optimum town and goes on to study the conditions of the second best optimum of land allocation models in a Thunen town [566] [598].

The fact that Thunen's model can explain all urban locations has not gone without question. Firstly, on empirical grounds [473] and secondly as Huriot has pointed out, on theoretical grounds, due to the contradictory role played by the central point in the Alonso town [616]. Indeed, the central point is in turn, a point where, the demand for goods and services, the supply of labour, the supply of goods and services and the demand for labour are concentrated according to whether we are considering commercial, industrial, or residential locations.

Nevertheless, the changing facets of Alonso's theory do not undermine the importance of the Thunen paradigm which Beckmann has recognized as a neo-classical theory of land allocation [465]. Moreover, he uses it as an analytical framework to study the residential area in a town [501] and the growth of a region in which all external trade is assumed to pass through a central town where industry and services are concentrated [500].

Furthermore, in urban economics, there is no longer any ambiguity in the role played by the center of the town in partial equilibrium models of residential areas. However, in all the analyses of choice and residential consumption after that by Alonso, the distance from the center is of less importance. Zoller, in an initial approach, builds a model of residential location which, as regards its form, resembles spatial interaction models [497]. It is based on the consumer's maximization of utility given a stochastic distribution of preferences. Senior criticized this analysis [546]. However, using the model's theoretical foundation, Zoller goes on to qualify the usual assumption of the probability of location diminishing with distance from the center. The author discovers a spatial friction effect which will be analysed later for him by Paelinck and Tack [689]. Finally because of its structure the model can be used in wider models of urban structures which simulate the residential and commercial occupations of urban space [334] [582] or optimize the flow of persons, locations and the spatial size of transportation networks [629]. Then, in a second approach Paelinck and Zoller re-examine the problem in a more suitable framework with the following characteristics: the indivisibility of residential consumption, durability of land allocations, irreversibility of location decisions over long periods, suboptimality of locations [655] [747]. They use discrete analysis within a framework of disequilibrium. This study is inspired by the theory of product differentiation and is level one from all the residential solutions, the others are at level zero. Just as in the oldest dynamic model, which we owe to Anas [498], the equilibrium conditions a la Thunen are not verified. Moreover, contrary to traditional theories, space is not redistributed instantly and totally in reply to variations in the parameters concerning the choice of residential location.

2. Weber's problem generalized

Weber had formulated the problem of the optimal location of the production unit in a restrictive way: he had assumed a "locational figure" reduced to three points, Euclidean distance, transportation costs proportional to weight and distance and fixed technical coefficients of production. Moreover, Weber's commentators and users have argued that the main criterion in optimal location is the minimization of total transportation costs. Nevertheless, Weber's work has remained at the heart of most advances in the theory of the location of the firm. This is why, since the nineteen fifties, a considerable amount of research has been devoted to generalizing the problem to which Weber's name remains attached. In the process the original assumptions have been progressively replaced by less restrictive ones. Hence Weber's model has been the point of reference for a major part of the theory of the location of the production unit.

In a seminal article published in 1958, Moses gave the first synthesis of the old theory of location and the modern theory of production [284]. In order to do that, the relationship between the point of minimum transportation cost and that of maximum profit had to be made quite clear. Isard had shown that these two points coincide when the technical coefficients of production are fixed [257]. Moreover he had moved on from "the locational triangle" to the case of any polygon. Moses asserts that this coincidence is improbable since it depends on the production function chosen. He focuses on the case of a function with variable technical coefficients.

Alonso carries this analysis further by introducing economies of scale, factor substitution and elastic demand into the Weberian model [364]. He concludes that the minimization of transportation costs is a dubious criterion and proposes an equivalence between the theory of location and that of rent. Not long after, Sakashita considers a linear and homogeneous production function and a demand function to the firm which is price elastic [393]. In one-dimension space, if the firm uses two inputs and if their transport costs are constant, since output transport costs are zero then the optimal location must be at an input place if the firm minimizes its total costs. This type of analysis is developed in several articles [444] [485]. Woodward, in particular, broadens Sakashita's theory and establishes under what conditions location is independent of the level of output [522]. However, Emerson [505], Nijkamp and Paelinck [514] strongly contest many of the conclusions reached by Alonso and Sakashita.

The problem is taken up again by Khalili, Mathur and Bodenhorn [524]. They go back to the Weberian "locational triangle" and the criteria of the minimization of production costs and those of the distribution of a fixed volume of output to the market. They establish three propositions: when a firm is constrained to remain at a given distance from the market, a necessary and sufficient condition for the optimal location of production to be invariable with the volume of output is that the production function is linear and homogeneous; the same necessary and sufficient condition holds true for the variable distance from the market; finally, if the production function is homogeneous, the firm comes closer to the market if there are increasing returns of scale, if they are decreasing the firm moves further away. Miller and Jensen widen the scope of this analysis by incorporating variable transport rates and by substituting profit maximization for the minimization of costs. In this way, they reduce the previous findings to particular cases [650]. Similarly, Mathur broadens Sakashita's analysis by studying the impact of constant transport costs [685]. Finally Eswaran, Kanemoto and Ryan develop a dual approach to the same problem in one and two dimension spaces [716]. Instead of using a production function, they use cost and profit functions which are duals of a particular production function. This method frees them from the assumption of a quasi-concave production function and enables them to study the case in which there are more than two inputs.

Thisse and Perreur adopt a wider analytical framework [627]. The firm uses more than two inputs and produces more than one output, with a production function which is independent of the location chosen. They come to a major conclusion. They make the most of a distinction first made by Alonso between the ex-ante minimization of transport costs and the ex-post minimization [364]. As in Weber's theory, the ex-ante cost is linked to the quantity of goods to be transported which is known a priori. The ex-post cost refers to quantities of goods determined a posteriori, once the location and production decisions have been taken simultaneously. In the case of variable technical coefficients, any point of maximum profit is always identified with an ex-post point of minimal transportation cost. From this general result, corollaries cover the particular cases (fixed or variable technical coefficients, price policies peculiar to different conditions of competition) which have been examined so many times by specialists since Weber and Predohl. Furthermore, this analysis remains valid even if we replace the minimization of transport costs by the maximization of the amenities offered by some places and introduce a further assumption about the monetary evaluation given to those amenities [740]. Hakimi's theorem, which proves that the optimal location of a firm in a transport network is either a market or a node [332], and its numerous extensions [377] [520] [726] [746] all hold true in the framework of the maximization of amenities.

One advantage of the Thisse and Perreur model is that it makes it possible to analyze the conditions of transport in greater depth. Indeed the properties of the point of total minimum transportation cost are still valid for the point of maximum profit. Hence, it follows that in any set of possible locations it suffices to determine the first point. We can then, according to the physical properties of the space considered and the technical conditions of the means of transport, use more appropriate metrics than Euclidean ones. This justifies the approaches adopted by Huriot and Perreur in which they use rectilinear metrics [511] [512], by Perreur and Thisse who use central metrics [541], by Thisse who uses a network-metrics [580] and various results presented by Witzgall [356]. Hansen, Perreur and Thisse also study the location of several units [700]. With a single norm, the solution to the problem is contained in the convex hull of the supply and destination points. Where there are several norms, the solution is to be found in the octogonal hull of those points. These findings generalize those obtained earlier by Wendell and Hurter for the location of a unit in the case of a single norm [521]. Furthermore, natural, human, technical and economic constraints can limit the optimization field, that is possibilities for the best location of the production plant. In this case, we find ourselves confronted by the constrained Weber problem [564] [719].

Moreover if two criteria are considered simultaneously, one being the minimization of the total cost of access, the other, the minimization of the maximal cost of access borne by the user, then we are back to the generalized Weber-Rawls problem. Hansen and Thisse have solved it for non-decreasing cost functions which are continuous with distance [721]. Hansen, Peeters, Richard and Thisse have solved it for the case where distances are represented by norms which can vary with the points under consideration, costs are increasing and continuous with distance and the possible locations belong to the union of a finite number of convex polygons [718]. Locations resulting from such economic calculations can be compared with voting processes. Hansen and Thisse show that they are identical if there is no cycle into the transport network and if the optimization criterion is the minimization of the sum of distances covered [722]. For a general network, the authors give the upper bounds for the ratio between the values of the objective functions which are attained at the two locations when, either the sum of the distances or the maximal distance covered is minimized, that is to say, according to whether one choses an efficiency criterion or an equity one.

More and more has been devoted to profit maximization where the choices as regards the location of the production plant, the inputs and the volume of outputs are interdependent, but without taking the theory of investment into consideration. It was Osleeb [539] and Osleeb and Cromley [621] who were the first to consider plant as a specialized and durable input in capital, which is variable in size. From this, Whitmore has established a link between the theory of location, the concept of time and the theory of capital formation [742].

The highly controversial works by Moses and Alonso engendered a considerable amount of analysis. At the same time, a new current of research led by Cooper emerged [316] [328]. Rather closer to operational research than to economic theory, it tackles the coupling of location and distribution problems. This coupling has given birth, in very specialized literature, to the Location-Allocation Problem [457] [503]. The allocation problem proper dates back to Hitchcock [119]. Distribution is to be optimized in a transport network which has several loading points and several delivery points. Cooper then combines Hitchcock's problem with the Location-Allocation Problem to create a new one called the Transportation-Location Problem [368] [470] [502] [588]. He made an original generalization of Weber's problem [384] which was to interest Tellier [495] and Beaumont [694]. Cooper [525] [634], Katz and Cooper [532] then widened the formulation to cover a random case. The destination points become random variables with given probability distributions. Wesolowsky introduces a rectangular metrics into this type of model [628]. Hence, progressively a stochastic Weberian model has been built (1).

It is interesting to note that only one part of his whole work, the theory of the point of minimum transport costs, is used to build the Weber paradigm. Isard is an exception to this rule, for he studies the possibilities and scope of using game theory to solve Weber's agglomeration problem [375]. In similar fashion, this has led Isard and Smith to apply game theory not only to Weber's theory but also that of Hotelling [376] [387] [388].

Alongside this development in the theoretical aspects of the generalized Weber problem there has been an increase in the number of operational models and algorithms proposed. Their history is necessarily dictated by that of theory and at times these two aspects of the same analysis overlap so much that they cannot be distinguished.

In the mathematical appendix to Weber's book [27] Georg Pick used by analogy, the Varignon mechanical system which could have led to confusion between Weber's minimum point and the center of gravity [517]. The following authors, Vergin and Rogers [380], Eilon, Watson-Gandy and Christofides [445] have all shown that the two points do not coincide.

At first iterative methods are used to tackle Weber's problem in its initial form [283] [313] [394] [567]. In similar fashion, Smith generalizes for any number of vertices, the determination of the Steiner point, found by Scott for a Weberian triangle [458], however Smith does not assume a homogeneous plane [739]. Hansen and Thisse develop an algorithm for the more complex case with the following characteristics: location is limited to the union of a finite number of convex polygons, distances are approximated by norms which can be different according to the points concerned and transportation costs are non-decreasing and continuous with respect to distance [723]. They then draw up a synthesis of the solution processes recommended whenever location must be optimized in a continuous part of a finite normed space [724]. Finally,

Hansen, Peeters and Thisse present an algorithm to solve the constrained Weber problem when the set of possible locations for the plant can be represented by a finite family of convex polygons [678]. The same authors solve the constrained Weber-Rawls problem by using an appropriate algorithm [720].

The models stemming from the Cooper tradition are the subject of associated algorithms in studies by Kuenne and Soland [487] and Juel [730].

At the same time, another line of research has reduced the solution of the generalized Weber problem to that of the Simple Plant Location Problem. The latter is, first of all, formulated separately by Stollsteimer [321], Balinski [325] and Manne [335]. The same constraints are used as in discrete locational models. Its objective is the minimization of total costs. To begin with, it is applied to the search for the Weber minimum point. Several heuristics have been put forward [481]. Erlenkotter has proposed a particular efficient algorithm [636]. Later on, the Simple Plant Location Problem is extended so that the search is now for the point of maximum profit. In addition to the variables already taken into account, the firm's optimal pricing policy now has to be determined. A specific model is associated with each pricing policy [611] [615] [698] [725]. Moreover, this type of model proves to be sufficiently flexible to incorporate uncertain demand [597], some procedures of the firm's sales policy [725] and the choice of the optimal production technique peculiar to each plant [615].

Finally, as MacKinnon shows [599] an algorithm to find the Kakutani fixed points can be applied to the generalized Weber problem as well as to that of Thunen and all spatial equilibrium models.

3. Hotelling's law

The article which Hotelling wrote in 1929 on the relationship between the formation of prices, market areas and the location of sellers was to have an enormous but belated impact [65]. The idea that spatial competition is responsible for locational clustering sparked off a wave of studies to test the validity of this important conclusion which had rather disquieting implications for welfare. In this way the result of the model became "Hotelling's Law". Little by little the debate around it became wider in scope and soon encompassed situations which moved further and further away from those initially covered by the author's hypotheses.

Hotelling had established this law of locational agglomeration in a duopoly situation and for a rectilinear, uniform, bounded market with perfectly inelastic demand. It was the first commentaries especially that by Smithies in 1941, which were to have the greatest influence. Indeed the law has sometimes been called the "Hotelling-Smithies Law" [127]. Yet Smithies had shown that the law only holds true under Hotelling's assumptions. With a linear demand function and variable transport costs the locations are optimal between the quartiles and the center.

Apart from these two cases no progress was made for twenty years. Then, very rapidly, the paradigm was built up: first of all several contributions modified the functional relationships between the elements and opened up the spatial framework of the initial analysis. Gradually this theoretical current tended to cover all the relationships between the individual locations and markets.

Stevens was to initiate a renewal in the theory of spatial duopoly when he introduced game theory into the field [302]. The Hotelling-Chamberlin model is assimilated to a two-person zero-sum game. Demand is assumed to be perfectly inelastic. This assumption was then replaced by those made by Lerner and Singer [97] and Smithies [127] and the results compared. At the same time, though independently, Jacot made the same attempt [319]. Linear programming which had been used extensively in game theory was then introduced into spatial theory [285]. A single analytical framework is used to study several different market structures and Jacot integrated the contributions by Lefeber [281], Goldman [279], Isard [280] and Isard and Schooler [290] as particular cases.

Then, Teitz, going back to the assumption of inelastic demand, introduced the possibility of multiple location for each duopolist, without any cost, and showed that locational equilibrium is only possible if each firm adopts a "maximin" location strategy [397]. Vickery adopted another topology to describe the market space and examined the locational competition of two identical firms on the circumference of a circle [342]. The advantage of this configuration is that it eliminates the problems connected with boundary phenomena, or the existence of exclusive hinterlands associated to the edges of a finite rectilinear market. As long as demand is elastic, firms scatter as much as possible at equilibrium, each seller is diametrically opposite the other. If demand is perfectly inelastic no equilibrium is determined. Devletoglou reaches similar conclusions when he adopts a two-dimensional spatial representation of the market, but he changes an assumption about the behavior of demand [349] [370]. When the difference between delivered prices is lower than a determined threshold, buyers show no preference for any particular seller. This condition is called "minimum sensible constraint of indifference". Moreover, the income-elasticity of demand is taken into account along with price elasticity and inventories are introduced. Under these assumptions, concentration is unstable. In general, firms scatter. Hotelling's law no longer holds true.

Gannon took these studies further by introducing the interdependence of strategies [475]. Indeed he considers conjunctural variations, that is to say, each seller's predictions as regards the degree of response from his competitor to an initiative taken by him. He uses an individual, general but well-behaved demand function, and a non-restrictive set of assumptions as regards behavioral response predicted by each firm vis-a-vis the other, and arrives at very original results. The concentration in the center depends on certain characteristics and predictions of the duopolists. Equilibrium can be reached anywhere on the market. Its realization at a precise point depends on the predicted locational responses which each firm attributes to its competitor. He then goes on to examine the optimization of market shares [507]. Finally, he establishes that Hotelling's law is valid for a set of perfectly elastic demand functions if the two competitors are interdependent in a non-Cournot manner [508]. On that particular point, Gannon continues the research which J.Hartwick and P.Hartwick have devoted to the problem of dynamic adjustment in spatial duopoly and its implications for the stability of locational equilibrium in the Hotelling-Smithies model [450].

With time this current of thought has become progressively more complex. To counter this Beckmann has decided to re-examine Cournot's oligopoly in a spatially homogeneous market where mill pricing is assumed [466]. He uses a simpler and simpler demand function, that is to say, a rectilinear one.

Two other main lines of development can be distinguished. Although related to the preceeding current of thought they stress particular problems and as such have given rise to rather specific works.

The first question tackled is the effect of differentiation on locational concentration. First raised by Lovell [429] this problem is taken up again by Kennedy and Copes [646]. They believe that product differentiation favours a Hotelling type concentration, in particular for hybrid market structures where both competition and monopoly are present. This conclusion was to be rejected by Eaton [472], Shaked [548] and Eaton and Lipsey [555] [674]. They establish the non-existence of equilibrium in a bounded spatial market and limit the centripetal effects of differentiation to local concentrations justified by the behavior of buyers who compare products at several points before purchasing. The same results are obtained by D'Aspremont, Jaskold-Gabszewicz and Thisse [673] [715].

Next, very specific research was initiated by the movement away two-seller models towards models with more than two sellers. Historically, the first of these works stemmed from another paradigm, Losch's theory of central places. Indeed Curry attempted to link Hotelling's law and Losch's theory when he pointed out the connections between the loschian location model and the Fibonacci system of numbers [401]. Scott returns to Hotelling's model and generalizes it by describing the process for entry into a branch [459]. Sellers adopt a short-run profit maximization attitude in a two-dimensional market. The new "entrants" enter one by one, at discrete time intervals, while other sellers can be pushed out by the newly arrived ones. Eaton, having examined free entry into a bounded market [472] go on to study the case of a market which is represented by a circumference and unbounded [589]. Multiple equilibria exist and, in general, the existence of zero profits is no longer an equilibrium condition as is the case for the single equilibrium solution with zero profits given by Telser [420] and Beckmann [440]. Finally, in a discrete market space Kuenne generalizes the Hotelling-Smithies model to cover any number of firms with specified functions [617]. By combining a non-linear program with a Weber minimum point algorithm, he derives the patterns of location and the social costs of oligopoly. Thus he extends Smithies' analysis [127] and so rejoins the works of Kuenne and Soland [487].

It is no exaggeration to say that of all the paradigms stemming from the history of spatial economic theories, Hotelling's law is by far the most controversial. Nevertheless it remains at the heart of nearly all the discussions about the relationships between locations and markets [696] [738] [741] [744]. In operational research it has given rise to the apologue of the "Ice Cream Man Problem" [516]. Finally it is known outside the field of economics.

4. The Christaller-Losch central places archetype

In its study of Weber's work the history of spatial economic theories has above all highlighted his point of minimum transport costs. As regards Losch, special importance has been given to his theory of economic regions. Although it was inspired by Christaller's central place model, it was built differently. Nevertheless scientists usually link the two and they make up the basis for a paradigm which is of considerable importance not only to spatial economic analysis but to theoretical and quantitative geography.

In the fifties and early sixties Christaller reformulated the general principles of his system which he then applied to some European regions [173] [315], to transportation problems [237] and tourism [327]. He also looked into the applicability of his model on a world-wide scale [294]. These studies completed the 1933 model [76] by providing him with additional theoretical details and empirical illustrations. At the time, they aroused a number of commentaries [303] [366] [400]. Then Alao, Nurudeen et al presented an overall presentation of Christaller's theory [606].

Theoretical geographers have also taken an interest in Christaller's central place theory and Losch's theory of regions. A number of publications by Berry and Garrison [276] [277], Berry and Pred [344], Berry [365] and Fano [405] set out to popularize and apply the two theories. These studies, as well as rather more abstract ones carried out by economists and geographers, have helped to turn the theory into an archetype.

During the sixties, Tinbergen was to present an original formulation of the spatial hierarchy of economic activities [305] [341] [399]. A comparison of his hierarchy and those established by Christaller and Losch has revealed different degrees of spatial integration in the three cases [659]. However, only Christaller's and Losch's theories were to be used to build the paradigm of central places.

First of all greater knowledge of this model required a close examination of its mathematical structures. In a long series of articles, Dacey reformulated its geometry, specified as one of finding the tesselation of Dirichlet regions in which packing density is maximized [329] [330] [331] [346] [347] [348] [358] [360] [385]. A Dirichlet primitive region is known to be a polygon with a lattice point at its center and such that any point on the plane is closer to that lattice point than any other lattice point. Thus Dacey introduces the concept of Brillouin regions used in studies on physical waves and cristal, to show that the hexagonal lattice is a consequence of the maximization of packing density. Finally Dacey studies hexagonal network deviations. A non-hexagonal and non-optimal model is possible, even in a homogeneous plane, as a result of institutional, environmental and human factors. This had led Dacey to introduce randomness into his formulation, since central places are located with a law of probability. On this point, his approach coincides with contemporary ones in which the central places system is viewed as the outcome of various stochastic processes [369] [378] [641].

From a different standpoint, Medvedkov has applied graph theory to describe the non-metric networks of central places [390]. This method has proved to be more general than the Nearest Neighbor Method used previously by Dacey [311]. Finally Haites has suggested that the theory of numbers should be applied to the system of numbers which Losch had associated with each agglomeration to express its rank in the hierarchy of central places [562] [592]. His suggestion has led to a heated exchange between Marshall and Beavon [618] [648] [631] [684] [665].

Alongside this research into the mathematical structures of the central place theory, another current of thought has explored its relationships with a long-established law, known as the Rank-Size Rule.

This purely empirical rule links the size of towns to their rank in the urban network hierarchy. It is quite an old law since it seems to have been formulated for the first time in 1913 by Auerbach [33]. It was taken up again by Lotka [50] and Singer who pointed out its affinity with a Pareto distribution [93]. Clark used it to explain the economic functions performed by towns according to their size [140]. Then the distance separating towns was introduced by Zipf who ended up with a gravitation formula [167] and by Stewart who related size to the spacing of towns [286]. Dziewonski has presented a synthesis [471].

In a seminal article which came out in 1958, Beckmann made the first attempt to reconcile the central place system and the Rank-Size Rule [275]. Dacey adopting a different approach made the same attempt not long after [359]. Following the criticism levelled by Parr [413], Beckmann and Mc Pherson presented a new version [422]. The debate was important in so far as the issue at hand was whether or not the central places theory could provide a logical basis to the empirical rank size distribution [433] [484]. Beguin has studied the conditions in which the Beckmann-McPherson model would be compatible with the Rank-Size Rule [666] [667], while Okabe has reformulated the law and introduced the expected Rank-Size Rule so as to make the law equivalent to the Pareto distribution [687].

Some commentators have noted that the Dacey and Beckmann-McPherson models share common points with the theory of the economic base which states that some urban economic activities are basic while others are non-basic. The former are exporting activities, they have substantial multiplication effects and they form the basis of the urban economy. The latter, having an intra-urban range of influence, are only residual. According to Dziewonski, the concept of urban economic base goes back to Sombart, who presented it for the first time in a study which was part of the second revised edition of "Der Moderne Kapitalismus" [371]. For Sombart the urban economic base is the most fundamental characteristic of urban development through the years. During the nineteen fifties, the theory of the economic base was the subject of a number of exchanges especially in the revue, Land Economics. It was then developed by Ullman [227], Sirkin [292] Ullman and Dacey [300]. Finally, in 1962 Tiebout was to give a comprehensive presentation of the theory [314].

Parr, Denike and Mulligan, having shown that the Dacey model is a particular case of the Beckmann-McPherson model, then established that it was compatible with the theory of the economic base [571]. In this way, along with the Rank-Size Rule, the economic base was in turn integrated into the central place archetype. Mulligan went further into the properties of such a hierarchical city-size model [686] [702] [732]. In particular, the connexion between this type of model and the input-output analysis which had been discovered not long before by Romanoff [545] was uncovered; moreover, the static comparative is introduced, in the manner by Beckmann and Schramm [467]. Finally, Horn and Prescott [643] then Haining [697] proposed methods to estimate the different models of the economic base and population multipliers.

Although the central places theory has absorbed earlier empirical models which strove to solve the same problems, some parts of it have been called into question.

First, the hexagonal market areas configuration was to be lengthily discussed. Losch had maintained that the equilibrium of such markets is compatible with the maximum number of firms per unit of area. In a seminal article, Mills and Lav claim that there is an error in this theorem [336]. The condition stating that the space must be fully covered by the networks of the market areas is not a necessary condition. In their view, the networks of circular areas are as stable at equilibrium as those of hexagonal areas and in the long run the profits of the firms are still positive. However, this theory has since been rejected by several separate commentators [425] [441] [468] [494] [510]. None the less, markets with a circular form have been shown to possess some of the equilibrium properties established in the case of markets represented by right segments [526] [552]. The solution to this problem depends on the effect on the market when new firms "enter". Indeed, Beckmann has shown that the shape of market areas, circular or hexagonal, is not determined by individual firms, but by the conditions of competition of all the independent firms [357]. Soon after, several authors analysed "entry" decisions as a continuous process and study the effects of this sequential phenomenon on zero profit spatial equilibria [590] [593]. From the point of view of welfare, the consequences are non-classical: Holahan and Schuler have established that multi-plant monopolies procure greater consumer surpluses and general welfare than those offered by competition [728]. Thus they have extended the analysis started ten years earlier by Beckmann in which he had maintained that for a spatial economy a producers' equilibrium can only be reached in a local monopolists' world and not in a perfectly competitive one [440].

The long debate about "entry" and its effects on location and market areas has opened up an analytical field common to two paradigms: Hotelling's law and the central places theory. In fact the discussions about "entry" in a generalized two-dimensional Hotelling model have tended to bring it back to a loschian type of spatial equilibrium.

Another line of research has called into question, not the hexagonal shape of market areas but their property of regularity. Losch had assumed that the agricultural production units in an area are uniformally distributed, self-sufficient farms. Such was the basis for his model of regular hexagonal market areas, the smallest area corresponding to the actual agricultural production unit, ie the individual farm. Losch then introduced into his analysis several economic, natural, human and political factors which all create a particular type of differentiation and distort the pure model obtained by taking into consideration the three basic factors of his model, namely, distance, the scale of production and competition. However, the resulting distortions which he had described, seem to have been forgotten. Indeed all the discussions concerning the reasons why the hexagons are distorted have focused on the pure model.

This line of research has studied the distorting effects of the non-uniformity of the economic environment. In his book published in 1956, Isard gives an inductive description of the effects urbanization can have on a homogeneous region [257]. The areas of the hexagons which are situated a long way away from the center must expand for the following reasons: production sites and the industrial population are to be found in smaller numbers; moreover one consequence of the industrial population differential is that agricultural production is less intensive and the rural population more scattered than in the central town's immediate hinterland. Sarly has formalized this analysis by reformulating Losch's production function such that the spatial agricultural economy can be taken into consideration in the case of a rural settlement [493]. Dacey in his examination of the mathematical structures of the central place theory has formally deducted the model's deformations which result from the systematic distortions of the lattice points according to the way in which the hypothetical probabilities are distributed [358] [360]. He goes on to analyse the measurements of the spacings obtained in the resulting model. Tobler [322] and Rushton [492] have reduced the problem to a map projection or more specifically to a map transformation which makes it possible to obtain the image of distorted spatial models as reflections of a single model. The reversed transformation gives back the spatial co-ordinates of central places. Tobler has transformed the locational co-ordinates so that equal areas have equal populations in the space thus produced. Using a different method, Rushton has estimated the expected distances between the points in the original space so that the areas surrounding the points have the same populations.

Finally, the analysis of the conditions of spatial competition and its effects on market configurations has led to research into demand, its definition and the role it plays in Losch's theory.

Losch believed that demand is more price elastic in space than it is at one point. Stevens and Rydell have studied whether or not it is in a monopolist's interest to absorb part of the delivery costs and they examine the possibilities and characteristics of a policy of spatial price discrimination [363]. They then compare the profitability of f.o.b. and uniform delivered pricing. The answer to these questions depends on the shape of the consumer's individual demand curve, more precisely, on its properties of convexity. In the same fashion, Gannon has made a deeper study of the demand properties in Losch's model [448]. Long has tackled the incompatibility of a regular model a la Losch with the existence of demand curves which are identical at any point in space [452]. The famous demand cones require special conditions.

Several authors have examined the relationship between the shape of the demand curve and the conditions in which the firm's f.o.b. price can rise or fall as a result of the spatial scattering of buyers and the "entry" of competitive firms [523] [560] [608] [632] [703]. Similarly, Carruthers has studied the connexion between the concentration of demand in ordered central places and the individual locations of production units [714]. Finally, a considerable number of specialized works have focused on all the relationships between models of the firm's spatial behavior and spatial pricing methods [477] [478] [561] [559] [563] [585] [613] [642]. Spatial behavioral models differ from each other by the assumptions made about spatial pricing methods. Maximum profit allocations depend on criteria such as total output, the consumer's total surplus or social welfare. Demand and cost functions, transport rates and market areas are taken to be exogeneous. Gronberg and Meyer, on the contrary, have built a spatial model of the firm in which the choice of the pricing method and the choice of the transport rate are taken to be simultaneous decisions, and as such they become endogenous variables [717].

As more and more factors and relationships have been taken into account so the central place theory has become progressively more complex. This is apparent by the number of attempts which have been made to study, in greater depth, one particular element of the theory for instance, the international aspects and the balance of payments [264], the determination of the location and the size of sets of subcenters inside the metropolitan areas [586], the role played by the cost of the infrastructure [437], the effects of the transportation costs of the products offered by one central place on its dependent area [427], or the importance of economies of scale both in production and consumption [647].

Section 2. NEW DIRECTIONS

In addition to the formation of certain paradigms, the contemporary history of spatial economic theories is characterized by new directions in research: the construction of models of spatial interaction; the development of the theory of general spatial equilibrium; the elaboration of a theory of spatial public economics, the birth of spatial econometrics and finally a deeper understanding of the concept of economic space.

Despite their common points with classical models each of the different currents of thought is largely autonomous as regards its own particular development.

1. The construction of models of spatial interaction

Since the nineteen fifties, both geographers and economists have been behind a new current of empirical and theoretical research in which gravitational physics has been applied to the analysis of human interactions in space. A new paradigm has been built which, unlike the previous ones, is not a legacy of the history of spatial economic theories.

However, the use of gravitation to explain human spatial interactions is not new. Carrothers has found that it was first formulated by H.C.Carrey in his work "The Principles of Social Science" which was published in the 19th century (Philadelphia, J.B.Lippincott and Co, 1858-59) [251]. It was directly inspired by Newton's law of universal gravitation (1680) which states that two bodies attract one another with a force which is proportional to the product of their mass and inversely proportional to the square of their distance. However, this formula has to be made more flexible when it is applied to human interactions. Gravitation is presented as an increasing function of mass and a decreasing function of distance. More precisely, two types of models have been used. On the one hand, gravity models proper which allow one to estimate the absolute number of interactions between two portions of space generally represented by points. The number varies directly according to a certain function of the masses associated with the points and varies inversely with a certain function of the distance separating the two points. On the other hand, models called potential models which aim at measuring the influence exerted by a set of masses on a unit of mass located at a given point in space.

The history of gravity and potential models applied to human spatial interactions revolves around discussions about the nature of those masses, the choice of a definition for distance, the calculation of the numeric value of its exponents and the evaluation of the gravitation constant.

At the end of the nineteenth century and during the first half of the twentieth century demographers used gravitational formulae. Masses are those populations whose migratory movements have to be explained. Zipf is representative of this first current which followed the "social physics" school of thought [129] [167]. Zipf measures the interactions between pairs of towns not only in terms of demographic magnitudes but also economic ones.

However, it was an earlier study restricted to the retail trade which was to serve as the reference model and point of departure for modern economic analysis of spatial interactions. Reilly presented it for the first time in 1929; then later in a version which has since become well-known [71]. The author sets forth an empirical law of gravitation which is applicable to shopping goods. According to this law, two towns attract the retail trade of an intermediate town to the neighborhood of the intersection point, in approximately direct ratio to the population of the two towns and inverse ratio to the square of the distances separating the towns from the intermediate town. The intermediate point where each of the two towns controls half of the sales is called "the fifty per cent point". It is the boundary point of the two reciprocal areas of influence. Converse provides the mathematical formula which is directly derived from the initial law [158]. This law, now complete, was later to be called the Reilly-Converse Law. Later discussions concern the applicability of the law inside the same town to determine the areas of influence of commercial centers [296] and the estimation of the exponent of distance and the gravitation constant [320] [337].

However, very rapidly, the law was to be given a more sophisticated wording. Stouffer [115] [299] and Strodtbeck [165] introduce the notion of intervening opportunities to express the attraction exerted by centers located between two points under consideration. Then Huff, using a probability model of relative potential, evaluates the potential market of a sales outlet [312] [318]. He makes the spatial behavior of the consumer more precise by considering that when a consumer decides to travel in order to purchase something, he is uncertain of how well-stocked the shops are. Finally this current of research focused on the problem of commercial locations [460]. As regards industrial location, Dunn tests the relevance of the gravity model when in order to determine the optimal location he attempts to combine two criteria: the maximization of the size of the potential market and the minimization of transportation costs [252]. Dunn gathers all the criteria together to form one criterion called "the location index". Finally, Hartwick calculates simultaneously the optimal location at the point of minimum transport cost and the market area where supply exactly meets demand [482].

Several contributions have introduced the time element into gravity models. Isard and Freutel suggest that the growth in a region's gross product during a period can be explained by the evolution in the income potential of the region relative to the whole system [189]. Moreover, the evolution in each region's income potential is used to build models to forecast the rate of growth in the regional population [278] [289]. Again, Warntz opposes the temporal income potential, taken as being representative of spatial demand potential with a temporal supply potential in order to explain variations in spatial price systems over time [274] [293]. Finally Isard tries to incorporate space and time into his analysis of spatial interactions by applying the general theory of relativity [451] [680]. It remains to be seen whether one can justify the passage from one analogy inspired by Newton to another borrowed from Einstein [453].

With time specialists in spatial interactions have questioned the basic analogy's capacity to provide explanations. So they have concentrated on finding a theoretical foundation. Learned conceptual models have replaced simple empirical ones. Two currents of thought can be distinguished: the current which bases the analysis of spatial interactions on probability theory and that which falls back on utility theory. In both cases, the link between the new theory of spatial interactions and physics is very loose.

It was the various attempts to find a theoretical justification for Stouffer's intervening opportunities which engendered the first probabilistic approach. Schneider set the movement going when he purported to determine the probability that an individual will move from a given place to another [291]. His study is based on the idea that this probability is a function of the number of opportunities, considered as the opportunities to stop, contained in the destination area. Harris has generalized this analysis [333]. Unlike Schneider, Harris assumes that movements are no longer made with the same purpose in mind but with very different ones. Furthermore, opportunities to stop have different weights according to what aim the individual is seeking. Behavior and decisions have gradually replaced mechanical attractions as the explanation to spatial interactions [579].

Another probabilistic approach has attempted to find a rational explanation at a macro-economic level by using the concept of entropy. Wilson derives a gravity model from the maximization of entropy under constraint. His findings, outlined earlier in a series of articles were published in a textbook in 1970 [438]. As this statistical gravity model has grown more complex so more and more technical research has become necessary [326] [538] [547] [569]. Certain researchers support this type of model [476] [557], while others have moved away from entropy and put forward probabilistic approaches using the Bayes theorem [406] or the properties of maximum likelihood estimators [446].

The other current of research has based its models on utility theory. The individual is no longer uncertain; he is described as a consumer of movements endowed with a utility function which depends on the number of interactions he has to carry out. Niedercorn and Bechdolt focus their analysis on the determination of the number of movements which provides the consumer with the maximal utility given the money and time he can devote to travelling [411] [432] [489]. A similar model is applied to movement of goods in a space composed of supply and demand points [537]. Mathur [430] and Allen [439] [462] who adopt a viewpoint a la Lancaster, consider a given type of movement as an input which gives rise to a set of characteristics, the outputs. Finally, Golob, Gustafson and Beckmann believe that the characteristics bear no relation to the type of movement but to the destination [509]. The economic agent under consideration is the household. The optimal number of movements made by a household, for a given reason and towards a given destination, is a direct function of the capacity of a destination to satisfy a given motive for a movement, and an inverse function of the cost in time and money borne by the household.

Although some supporters of entropy have dismissed the notion of a utility function [574], the two schools have been reconciled on occasions [480] [577] [551]. Maximizing entropy under constraints and maximizing utility under a budgetary constraint are in fact analogous problems [443]. Similarly, a certain synthesis within the same current is possible [565] [578]. It appears that in all cases, the fundamental problems are those relating to the amount of information which economic agents have about the properties of sales points or destination areas. Gould has noticed that mental maps correspond, in some available experiments, to the objective maps obtained from a potential formula [558]. In the same fashion, Cadwallader underlines the importance of the way in which individuals appreciate the elements situated in the objective space [550]. However, it is Fustier, in a rather different approach, who introduces an appropriate analytical framework to the study of spatial interactions in a subjective and hence imprecise space

[676] [695]. By using fuzzy subsets theory he can formalize, in a rigorous manner, the concepts of imprecise attractions and fuzzy attraction areas. In this way, the human aspect is no longer such a barrier to the mathematical expression of spatial interactions as it was in traditional formulations.

2. The development of the theory of general spatial equilibrium

Losch formulated a number of equations which expressed the conditions for a general spatial equilibrium [112]. However he did not probe further in that direction. He preferred to devote his energy to a detailed elaboration of a theory of economic regions which he believed to be the necessary link between theory of the location of an individual plant and the theory of general spatial interdependence. This led his successors to concentrate on the central places theory to the detriment of the theory of general spatial equilibrium. Losch's equations have aroused few comments if not those made by Beckmann in 1955 [234] and the strong criticism from Mougeot twenty years later [568].

The theory of general spatial equilibrium is characterized both in its origins and development by great diversity and dispersion.

The first contributions dating back to Enke [184] and Samuelson [211] gave birth to so-called models of spatial equilibrium of separated markets [272] [338] [435].

Elsewhere, in an important article published in 1957, Koopmans and Beckmann study the optimal allocation of a finite number of firms to the same number of sites which maximizes aggregate profitability given the costs for the transportation of intermediate goods between each pair of firms [271]. These intermediate flows create an interdependence in the location model. One particularly contested conclusion is that a decentralized market economy cannot generate land prices for each site which guarantee a social optimum. The commentaries which have followed either relax the indivisibility assumption [431] examine the model [483] [594], question the conclusion described as guesswork [530] or simply reject it [731].

The progressive elaboration of a theory of general spatial equilibrium has stemmed from a desire to convert all the existing spatial theories into particular cases of a more general theory. This line of action characterizes the early works by Isard all brought together in his book published in 1956 [257].

Lefeber, on the contrary, in an original work which did not arouse as much attention as it deserved, purported to elaborate a theory of general spatial equilibrium on the lines of modern neo-classical analysis, using the same method, that of the Lagrange multipliers [281] [282]. Moreover, he discards the assumption of a homogeneous plane a la Losch and a la Isard and considers a finite number of production and consumption points. His other postulates are classical: a competitive market, convex sets and concave functions. From this basis he elaborates the theory by progressively relaxing certain complementary assumptions. However the author centers his whole argument on a two-point space, two sources of factor supply and two products. It is only in a short appendix that he attempts to generalize the model to encompass any number of factors, products and points of consumption and production. Goldman also uses economic programs to deal with a model in which there are three points of production and consumption, situated at three sources of raw materials [279].

In his work, Boventer shows, that the desire for a sufficiently general framework into which the existing theories of Thunen, Christaller and Losch can be integrated, has become all-important [308] [309] [310]. The assumptions he makes, for instance, incomplete competition, are sufficiently flexible for the following elements to figure in the general equilibrium described: transportation costs are taken into account, individual production functions are not necessarily linear and homogeneous, the positive and negative external economies are integrated, maximum profit is not the sole objective for businessmen, the capital elements necessary for production are incorporated, inferior consumer goods and the "Giffen-effect" are, if the case arises, introduced and finally, the analysis encompasses administrative activity. While it is true that the conditions for the existence of general spatial equilibrium are established, those for the stability of the system are not.

Several authors have since attempted to combine the various traditional theories. They use either, activity analysis like Beckmann [381], the general theory of systems like Isard et al [407] or analytical models like Serck-Hanssen [434] and Greenhut [426]. Furthermore, externalities and environmental phenomena have also been analyzed spatially [486] [506] [620].

Finally, Takayama and Judge on the one hand [339] [340] [396] [461] and Mougeot on the other [568] [652] have made independent but more or less simultaneous attempts to introduce space into welfare economics. Using a concave program (2) these authors show that the well-known equivalence between optimum and equilibrium in non-spatial normative economics also holds true in a spatial economy. The two approaches differ in their treatment of transport. Takayama and Judge introduce transport services as elements of the transformation functions of production firms. Mougeot considers carriers endowed with their own production function and a specific market for transport services. Moreover, his model has been completed by Chevailler who introduces external effects into the analysis [587].

As long as the theory of general spatial equilibrium has as role to unify the existing models then it suffices to define a wide enough set of assumptions and choose a relevant analytical tool. However, when general spatial equilibrium theory aims at introducing the spatial factor into modern theories of general equilibrium and social optimum then the tools of the latter prove to be restrictive and not very appropriate [568] [587].

3. The elaboration of a theory of spatial public economics

The origins of spatial public economics do not go back very far in history.

For a long time, authors when exploring action by the State and public bodies relative to economic space, merely opened up the debate on the aims, means and effects of public intervention. Indeed these developments came in addition to theoretical analysis but were never really integrated into it. Losch adopted this attitude, for example. In his theory of regions, when he posed the problem of the relationship between political and economic territories and then described the distorsions in the haxagonal pattern as a result of State intervention. In the same manner, when in his trade theory he studied how product transfers and factor combinations are regulated [112]. A great deal of other similar examples could be given. The fact that partial or general equilibria do not meet certain social optimum criteria has always nurtured reflection on the role played by public authorities. The only exception to that is the elaboration of a normative theory in France. Indeed, Allais, as early as 1943 was to lay the foundations of the spatialization of welfare [134]. He underlines the role played by rent and the influence exerted by the plurality of markets on the conditions of social return. However, Allais who initiated marginal cost pricing in public firms, did not find much support from later authors. He receives partial support from Oort for example as regards optimum transport prices, where distance necessarily plays a role [412].

Spatial public economics came to life with a contribution by Tiebout in 1956 [266]. Two years earlier, Samuelson had presented his famous concept of a pure public good. Tiebout introduced space into the theory and created the concept of a local public good. The supply of public goods is not uniform in space and that non-uniformity could provide a solution to the problem of the revelation of preference and of the demand for such goods. Tiebout examines a set of localities characterized by tax systems and supplies of public services. Individuals with preferences which take the elements of these policies into account divide themselves up amongst the localities so as to maximize their utilities. Thus individual preferences for public goods are revealed by the choice of a locality, that is by applying the principle of "voting with one's feet". Public expenditure is allocated efficiently at equilibrium. As Thisse and Zoller remark, Tiebout's contribution is an ingenious extension of Samuelson's theory [752]. Any public good can be consumed by all those who are in the locality where it is supplied. No-one outside that locality can consume it. Supply areas do not overlap. Since movements inside a given area are assumed to be costless and movements outside are assumed to be very costly, it follows that the localities are considered to be points without any reciprocal interaction. Finally, Tiebout determines a simple partition of all the individuals into homogeneous groups.

In a second article in 1961, Tiebout reversed the problem [304] and adopted a formulation which Smolensky, Burton and Tideman were to develop several years later [436]. Users of public services are distributed in space at fixed locations such that their utilities diminish according to the growing distances separating them from the nearest public goods supply points. Thus space introduces some types of exclusion in the consumption of public goods. Henceforth, local public goods no longer possess the property of purity as Samuelson's public goods did. The result was that the need for a specific theory of local public goods became apparent and engendered several currents of research.

First of all as regards the Tiebout theory itself, research has gone into the role played by distance according to whether local public goods are transportable or not [683], and into the form of the demand function for services offered at given fixed points [707]. Several empirical and theoretical studies have been conducted [705] and the theory has been examined critically [612] [657] [708] [713].

At the same time, once the interdependence between the location of public facilities and the mechanism for their allocation in space had been established, researchers tended to focus their attention on the optimal location for public services. Operational research has prevailed since Teitz maintained in an article that several location models used for private firm could be applied to public enterprises [398]. This is how the first models of public facility location appeared. The social utility functions used are rather particular. They are in fact the sum of all consumers' utilities, or more often than not, the sum of the individual travel costs taken to be a measurement of the disutility resulting from dispersion in space. Yet, using these optimization criteria tends to foster inequality in the conditions of accessibility to local public goods [619]. For this reason, other researchers have opted to use the maximization of the most disadvantaged consumer's utility as their criterion and have as a result rejoined the location-allocation models based on the Rawlsian equity criterion. In an effort to go beyond the opposition between equity and efficiency some authors have endeavoured to determine the Pareto-optimum configurations while leaving the planner to be guided by his preferences when chosing between equity and efficiency. All these models have been the subject of critical inquiry [661] and syntheses [699] [748].

From there the theory of local public goods moved towards works encompassing a theory of local public economics. That implied considering local public goods within general equilibrium and optimum theory [528] [575] [614] [639] [662] [675] [677] [712]. Little by little the different problems in question have been defined and Deloche has proposed a classification [743]. A public good's range of influence can either be identical to that of the authority which decides to supply and finance it or it can be different. In the first case, if the good is produced by the State, it is a pure public good. If it is produced by local authorities it is a local public good, endowed with the possibility of exclusion and subject to congestion. In the second case, two possibilities must be distinguished: either the public good's area of influence is contained within the decision-taking authority's area of influence or vice versa. In the former case the beneficiaries view the good as being both public and pure while for the local authorities it is private. In the latter case the good is always public and pure. Each of these situations corresponds to specific problems but their solutions have not all been elaborated to the same degree.

4. The birth of spatial econometrics

The first specialists to show interest in the problems posed by the analysis of regionalized variables were statisticians [352] and quantitative geographers [361] [383] [553] [668]. The interdependences which generally exist between the measurements of regionalized variables both reflect the non-neutrality of space and question the use of traditional statistical methods.

The birth of spatial econometrics proper is more recent. According to Paelinck and Klaassen it was germinating in 1966-67 [688]. The first paper explicitly under that title dates back to 1974 [595].

The specification of spatial econometric models is obviously based on the elements of a spatial economic theory. Since there are so many reciprocating influences between regionalized variables, whether relating to consumption, production, investment, etc., spatial models have been formulated in an interdependent way. In addition, more often than not, those spatial relationships are asymmetrical.

Furthermore, spatial econometrics has had to take into account the so-called principle of "allotopy" which states that economic phenomena situated in a given space can be explained by factors located in other distinct spaces. In general, the influence exerted by the level of spatial aggregation poses rather awkward problems.

Another difficulty inherent to spatial econometrics lies in the fact that the ex post interaction between variables differs from the ex ante interaction. Indeed anticipated flows differ from actual ones which depend on factors other than the points of origin and destination of those flows.

Finally, space viewed as containing economic activities is bi-dimensional and should be understood as such; it therefore proves irrelevant to use punctual models in which points replace surfaces.

All these characteristics make the problem of specification all the more original in spatial econometrics [709]. The same applies to the problems of identification, estimation and test [596] [633] [653] [679] [688].

The latest developments cover the analysis of frequency distributions over space [681] [709] and the modelization of regional disequilibria [654] [663] [664] [710].

5. Furthering the concept of economic space

Until now authors have concentrated on analysing economic phenomena in a space considered as a container, or what comes to the same thing as a support. Strangely enough they have hardly even attempted to find out what exactly economic space is.

However since the end of the 1960's one current of research has endeavored to explore, in a systematic way, the different structures of economic space, to analyse their roles and bring out the effects they have on economic mechanisms and behaviors.

From Thunen to Losch right up to contemporary authors most of the great theoretical reference models combine economic space with a favored mathematical model: that of an isotropic, homogeneous, convex space endowed with euclidean distance [372]. With this formal representation go a number of restrictive assumptions: transportation costs which do not differ with direction, equal possibilities for movement anywhere and equal conditions of accessibility, consumption and production conditions which are similar at all points, etc.

Furthering the concept of economic space has consisted in diversifying the mathematical representations of space so as to formalize its structures in the least restrictive way. The properties confered on economic space by the mathematical structure are also made explicit. In other words, each type of problem is endowed with the appropriate analytical tools. Instead of restricting the analysis to the particular case of a convex subset of one or two dimensional euclidean space, the appropriate mathematical structures will be used after a great deal of thought.

Generally speaking, an economic space is a finite or non-finite set of points, equipped with characteristics and a configuration. Configuration is taken to mean the position of points with respect to each other, the distance separating the points, the form and the size of their set and subsets [706].

The question of positional relationships falls within the province of graph theory which was introduced by Ponsard to formalize a discrete and non-metric space. He uses it to deal with the optimal location of the production plant in a non-weberian space [362]. The space is not necessarily isotropic and homogeneous. Its accessibility depends on the associated graph's properties of connexity. Having been formulated for the case of competition this models was extended by Scharlig to cover monopolistic competition [418] and by Gadreau to cover oligopoly [447] [474]. Then Guigou went into the possibilities of combining this approach with multicriteria choice models [449]. In the same vein operational solutions have been found to the problems of the optimal location for warehouses [464] [518]. Finally Ponsard [454] and Bounon [524] have gone from the location of a plant to that of a branch of activity.

Graph theory has proved relevant when handling accessibility inside a spatial structure such as that of hierarchical models, and in particular that of central places [392], or that of gravitational type interactions [456] [491]

Transfer graphs (or signal-flow graphs) have been found to be particularly interesting. They are associated with simultaneous linear equation systems. The elements of the set of points are the dependent and independent variables; the mapping is given for the relationships between the variables; the parameters of the equations are used to evaluate the arcs of the graph. This tool, which proves well-adapted to the analysis of intersectoral and interregional flows has been used by Ponsard to build a model of interregional economic equilibrium [414]. The matrix algebra, which is quite commonly applied to this type of analysis makes it possible to study many properties; with transfer graphs new properties associated with the positional relationship of the regions can be discovered. Several authors have used this tool to further the analysis of dominance phenomena and to assess its relevance to other problems [490].

In short, graph theory, the successor to "Analysis Situ", provides the situational geometry which is essential to formalize the positional relationships between points in a given space.

Next, approximating the distances separating different loci implies chosing a metric with which to endow the space of the points representing those loci [690]. There is no justification for the exclusive use of euclidean distance. The rectilinear distance between two points, which is measured parallel to the axes of an orthonormal system, is easier to use analytically and can prove more flexible when representing concrete movements in several cases. Gambini, Huff and Jenks [386], Beckmann [442] and Fustier [556] all use rectilinear distance in their studies of market areas, as do Huriot and Perreur [511] [512] in their location models. Finally oblilinear distance between two points, which is measured parallel to the axes of a non-orthogonal system, is more general than rectilinear distance. Perreur uses it to deal with the optimal location of the firm in the case of a multi-plant concern [540]. In all these cases, the assumptions of the isotropy, homogeneity and accessibility of space cease to be a constraint since an infinite number of possible theoretical paths between two points exist. It is always possible to find at least one satisfactory approximation of the real distance. Moreover separate expressions of those distances can be given for each axe of co-ordinates such that two transportation charges can be introduced.

Another family of metrics, central metrics, is appropriate when evaluating real distances in a nodal structured space. The transportation network is then organized around a central point. Either the paths converge towards it (radial metric), go round it (peripheral metric) or these two means of travelling are combined and then a choice made (circumradial metric) in which the shortest of the radial and peripheral paths is preferred. Perreur and Thisse have used this circumradial metric to deal with such diverse problems as the optimal location of the firm [541] and urban travels [542], but in both cases, spaces are structured around a central point. The main advantage of the circumradial metric is that it discards the assumption of an isotropic space since different costs of movement can be attributed to the radii and to the peripheries.

Finally when the properties of a topological distance prove to be too restrictive then the definition can be weakened and weak metrics used, as Rouget has shown in the description of some urban spaces [602].

Any study of spatial economic forms naturally implies analysing its topological properties. It is strange that all through the history of spatial economic theories, authors have shown little interest in topology, which means etymologically the science of loci. Thunen's or Alonso's concentric rings, Weber's triangle or Losch's nested hexagons etc, are elementary geometric representations which not make it possible to formalize the representations of universes having any forms whatever. Ponsard, in his reconsideration of the theory of market areas has shown that there is no close interdependence between the surface and forms of those areas [544]. First of all he examines the metric properties of the market areas and shows that the particular topologies are inferred from the distances chosen. The form of the isovectors and isodapanes depends on the metric adopted. Then the study goes on to examine the purely topological properties of those areas viewed as discs. The variations in price and transportation costs give rise to elastic distortions in the plane which leave the non-metric properties invariant under the conditions of continuity. Hence an infinite number of market areas exist with different forms which are topologically equivalent to the discs. Arpin persues this line of analysis when she studies the topological properties of the graph representing an hierachical system of central places [607]. Similarly, Rouget applies this type of analysis when examining the topological representations of urban space [576] [692].

Thus graph theory as well as metric and topological space analysis have provided spatial economic theory with several relevant analytical tools [573].

However, in addition to the formalized description of the configuration, spatial analysis must also go into the properties with which its component localities are endowed.

First of all economic space is endowed with physical properties related to the soil [531]. There is a limited amount, hence it is a scarce resource. The quality of the land varies with its position. Some specialists in land economics have taken it upon themselves to consider space not only as a container or a support but as an economic good. Guigou stresses the functions it fulfills [479]. Space is both a producer good in agricultural, industrial and service activities and a consumer good to households for their accommodation and leisure. Furthermore space can be appropriated and fulfills the functions of a capital good as well as those of a speculative instrument. The approach first used by Guigou to study transformations in agricultural space has lead to a more general consideration of all locations as a result of land allocation. Huriot has formalized the analysis of the land use transformation process by using a recurrent programming model [616].

Having broken space down into finite-sized zones characterized by their positions, he observes the allocations effected and the expected returns on potential allocations. He can then show how exogeneous factors result in differential growth for the various activities, disturb the equilibrium obtained and tend to create a new state of equilibrium. The objective function is the maximization of the total net surplus obtained from the land use. The constraints cover the limitation of the available surface and the relative stringency with which land is allocated. The solution to a program for a period gives the surfaces allocated to the different activities in each zone. For two successive periods the solution to the first program falls into the second program's constraints. Finally the dual variables can be described as scarcity rents. This recurrent model which describes a sequence of disequilibria is far removed from the Thunen paradigm which describes a static equilibrium in a radio-concentric space. Land allocation is sub-optimal at any point in time because the mobility of activities is imperfect [644].

Next, economic space possesses geographical properties relating to the environmental characteristics. Scheubel has incorporated then as well as the traditional factor of distance into the spatial theory of the consumer [626].

Finally, more generally economic space is equipped with several material and immaterial characteristics which have been studied by using multi-dimensional analysis. Rouget has examined the spatial equilibrium of the consumer in the analytical framework of the lancasterian theory of choice [736]. The consumer's preferences are directed to all the characteristics of the located goods and not to all the goods themselves. In the case of current goods, those characteristics relate to the properties of the goods as well as their relative positions with regard to the consumer. Distance which is viewed as a negative utility is thus a particular characteristic. If the choice bears on the space itself, viewed as a consumer good, the author is then back to the residential location problem and provides a solution by using the core concept of a preference graph which draws on multi-criterion choice methods.

Up to now spatial analysis of the consumer has aroused few contributions apart the generalization of Slutsky's equations by Long [452] and Zoller [581]. Prices depend on location and movements in space give rise to substitution and income effects. Moreover, most of the research carried out has focused on residential consumer behavior. Papageorgiou even goes so far as to ask if the priorities haven't been reversed in that the theory of urban residential space presupposes a spatial theory of the consumer [600].

A consequence of the furthering of the concept of economic space is that a theory of fuzzy spaces has been elaborated.

Since it first began spatial economic analysis has been restricted to describing precise spaces. A precise space is one which has or has not given characteristics. Economic agents located there behave in a coherent manner and make exact economic calculations. This type of analysis is based on a binary logic; the presence or absence of spatial characteristics; preference or non-preference of agents with regard to possible actions.

Ponsard has introduced the concept of imprecision into spatial analysis and developed a formalized theory of fuzzy economic space. The mathematical tool used is the theory of fuzzy subsets [572] in association with n-ary logics.

At first this research was applied to regional analysis [623] in order to describe spaces which "more or less" possess given characteristics. The fuzziness affecting those spaces is expressed by their variable degrees of membership to given sets of characteristics. An economic region is indeed a typical fuzzy space. It possesses, to differing degrees, the characteristics of an industrial, agricultural, tertiary, etc, space. It is more or less homogeneous, there is no clear-cut frontier. This way of viewing the region has been associated with an appropriate method of numerical taxinomy devised by Deloche [609] and Tranqui [660]. It is original in that it retains all the available information and does not partition space on the basis of arbitrary separation criteria or thresholds. Thanks to this method several spatial subdivisions, which may be disjointed or not, can be obtained according to the degree of homogeneity required, such that regional frontiers are not unique, but depend on the heterogeneity of the content of the territorial units which they surround. This method has been applied to the regionalization of the French economy by Tranqui [660] and to the European economy by Ponsard and Tranqui [751]. Interregional relations are similarly imprecise in that the relationships as regards influence and dominance in a hierarchy of central places are fuzzy. Not only are these relationships diffuse but they are in part reciprocal. They constitute a highly entangled web. So as to fully appreciate these complex relationships Ponsard has elaborated the theory of phi-fuzzy graphs and used it to analyse a system of central places and its hierarchy [624]. De Mesnard, in the same vein, has applied this type of analysis to a table of interindustrial and interregional trade [635]. Furthermore, Ponsard using the theory of fuzzy systems has described how a regional economy functions dynamically [656]. Thus regardless of whether urban games exist, fuzzy regional games can be devised by simulating human behaviors and their relations inside the regional economies. Economic spaces other than the region are vague in character. This is true of the zone of attraction of a commercial unit to which Fustier applied the theory of fuzzy subsets [638] before extending it to the set of all attraction and interaction phenomena in economic space [676] [695]. Similarly, in an urban economy, Rouget has shown how fuzzy topology makes it possible to analyse formally the images residents draw up of the town in which they live [576] [737]. These mental maps of the town are all the more important since they enable urban agents' motivations to be explained.

With time, in addition to these fuzzy spatial analyses, a theory of the imprecise spatial behavior of economic agents has been developed. This new line of research answers the need to take the analysis further and so move away from a description of fuzzy universes towards an explanation of the human behaviors which shape them. First of all, Ponsard built a theory of value in a fuzzy context [691] [733]. The individual whose information is imperfect and ability to discriminate between possible actions limited, has only one imprecise preference-indifference structure which can be represented by a fuzzy total preorder. Under certain conditions, it is possible to represent the imprecise preference numerically which has led to the construction of a fuzzy utility function. The analysis then moved onto the optimal demand and fuzzy spatial equilibrium of the consumer [734] and the optimal supply and fuzzy spatial equilibrium of the producer [749]. From there it is quite easy to show that these models of economic behavior are particular specifications of a general optimization model of a fuzzy objective function under elastic constraints [735]. Finally, when there is a precise objective but the constraints are fuzzy a simpler economic calculation can be proposed [750].

So little by little researchers have elaborated a theory of fuzzy economic spaces which is axiomatically general in nature and which embodies ordinary theory as a particular case [706] . Indeed just as a certain event is the limiting case of a random event, so the precise and random events are limiting cases of a fuzzy one.

CONCLUSION

As its paradigms have been elaborated and new directions developed so spatial economics has consolidated and extended its field of analysis.

Greater depth has been added to the analytical foundations. Beguin and Thisse were the first to put forward an axiomatic approach which borrows concepts from the theories of metric spaces and measure [669]. However, the set of axioms presented is restricted to the most traditional representations of economic space, ie. those in which the spaces described are endowed with a distance and precise characteristics. This approach disregards those formalizations using graph theory or general topology which both make it possible to describe space irrespective of any metric property. Similarly it does not take into account fuzzy economic spatial theory in which spatial economic characteristics are imprecise and are not, strictly speaking, measurable [706]. Spatial economic analysis as it now stands, rests on a less restrictive set of axioms.

Nevertheless, it cannot be claimed that research has extended uniformally to cover all the different fields of economic analysis.

Indeed such important theoretical fields as money, macro-economics and income distribution have been almost overlooked. Similarly there has been no systematic study of the relationship between spatial economics and demography. Spatial research, from Termote's original contribution [421] to the works by Alonso [630], focuses on population migrations, what causes them, how they work and their effects.

As yet, specialists have shown little interest in the introduction of the time element. Webber has made the main approach by introducing uncertainty into spatial models [496]. Uncertainty is analysed for its effects on land use, regional growth, the establishment of towns and decision-making as well as for the costs which are associated with it. The impact of innovation and education on the locations is also examined. Several contributions to spatial dynamics have been made concerning regional growth [419] [543], economic planning over space [513] and spatial systems [649].

Thus, in the second half of the 20th century, vast scope remains open to spatial economic theory. Nevertheless, spatial theory has already developed to such an extent that it has been possible to extend and revive the use of applied spatial analysis in a wide range of fields, for instance, marketing and market surveys, the setting up of firms and public services, the location of crops, land problems, input-output analysis, regional and urban economics, transportation policies and national and regional planning.

---

(1) A full bibliography of publications by Leon Cooper and the Doctoral Dissertations he supervised was brought out after his death in 1980, in the Journal of Regional Science, 20, 1980, 525-553.
(2) A special issue of the journal, Regional Science and Urban Economics, is devoted to the quadratic models of spatial allocation. See: Quadratic Models of Spatial Allocation, R.S.U.E., 8, 1978, 1-116.

Conclusion

The History of spatial economic theories opens in the 18th century. To this day, in the second half of the twentieth century, it still has not reached its close. For that reason a book such as this has no real conclusion.

Nevertheless, this History does arouse the need to reflect particularly deeply on the present state of economics and its relations with space and time.

The effects of the break between spatial economic analysis and the Classical English School are still being felt with the result that economics now finds itself in a paradoxical situation. On the one hand, owing to the fact that commonly acknowledged economic theory continues to ignore the spatial element, the laws it sets forth necessarily lack the general significance that this economic theory should confer on them. It must be stressed that as a rule, these theories only hold true in a punctiform economy, that is in an economy where all the agents and goods are concentrated in one single point. On the other hand, spatial analysis remains very much the domain of specialists. Indeed most non-specialists are totally unaware of the analytical frameworks which have been developed, the concepts used and the findings reached. Such ignorance could be justified if spatial theories were simply deduced from non spatial ones by merely introducing additional appropriate assumptions. However that is not the case. A law which holds true at one point does not necessarily do so at a finite or infinite number of points.

The History of spatial economic theories follows the slow progressive identification of the configurations which characterize the spatial organization of agents and goods. All the significant advances in this field, the milestones in the History, have been associated with a formalized analysis of a new configuration. This is as true of the precursors' simple representations, Thunen's rings, Weber's triangle, Hotelling's right segment, Losch's hexagons as it is of the highly sophisticated structures which some modern authors have explored.

With so many formal structures representative of space, spatial economic analysis is inevitably more complex than dynamic analysis.

Time is "measurable", as used in the mathematical theory of measure. Furthermore both an arbitrary system of measure units and precise measurement tools exist. It follows, therefore, that time possesses the property of being totally ordered by the inclusion relation. It can be expressed in terms of intervals, which may be infinitely small, with the help of units which are multiples of the smallest unit taken from the measure system selected. The uniqueness of the formal time structure is in no way belied by duration which is so often contrasted with chronological time. Any duration can be projected on the chronological time axis.

Space, on the contrary, is not always measurable. Spaces observed do not necessarily bear a relation to a simple dimensional space, such as tridimensional space for instance. Recent history of spatial economic theory has shown that more diversified ways of representing space are now being used which require stronger or weaker formal structures. Spatial analysis is confronted with the problem of which model to choose. That choice is a very difficult one to make given that the different formal structures are not neutral. To adopt a formal model, a priori, without first investigating the question thoroughly might well limit the analysis to a particular case or even place it in an irrelevant space. Moreover, the findings reached in the framework of the abstract model chosen could not be extended to other abstract models unless some rigorous proof was provided.

Thus time seems poorer than space when their formal structures are compared. As a consequence spatial economic analysis presents specific difficulties which have yet to be overcome. Finally, any scientist on examining the numerous formal structures of economic space and their ensuing properties must realize that it is no easy task to generalize findings beyond the universe for which they were established, whether that universe is the point or specific configuration of points.

EASTER 1982.

Mathematical Appendix

I - VON THÜNEN'S MODELS.

1 - *Concentric Rings*.

A - The formula for the series of local prices as a function of distance x.

The problem is to compute the value of rye on an estate located x miles from the market.

If a vehicle transports a weight of 2,400L, or $\frac{2,400}{84}$ S, and the horses need 150L of forage to be able to cover a distance of 5 miles, or 30xL per x miles, then $\frac{2,400 - 30x}{84}$ S will be sold in the market (1).

At a price of 1.50T, the seller will receive (2):

$$(\frac{2,400 - 30x}{84} S) \cdot 1.50T = \frac{3,600 - 45x}{84} T$$

Since the transportation for a 5 mile distance costs: 2.57S + 1.63T, and for x miles (3):

$$\frac{2.57xS + 1.63xT}{5}$$

we obtain:

$$\frac{3,600 - 45x}{84} T - \frac{1.63}{5} T - \frac{2.57x}{5} S$$

or:

$$\frac{18,000 - 361.92x}{420} T - \frac{2.57x}{5} S$$

an expression of the net receipts from selling the quantity $(\frac{2,400 - 30x}{84} S)$ in the city.

Thus:

$$\frac{2,400 - 30x}{84} S = \frac{18,000 - 361.92x}{420} T - \frac{2.57x}{5} S$$

or:

$$\frac{12,000 + 65.88x}{420} S = \frac{18,000 - 361.92x}{420} T$$

---

(1) L, Hambourg pound, unit of weight
S, Berlin's Scheffel, unit of measurement of grain
1S = 84L.
(2) T, ordinary Thaler, monetary unit.
(3) Thünen adds Scheffels and Thalers (that is, a physical unit of measurement with a monetary unit). This formal inconsistancy seems to originate from the need to isolate the real elements from the monetary elements, according to the data of that time.

And 1 S is worth : $\frac{18,000 - 361.92x}{12,000 + 65.88x}$ T

which can be simplified to : $\frac{273 - 5.5x}{182 + x}$ T .

B - The formula for the series of transport costs as a function of distance x.

The requirement is 30x of forage for each full vehicle, that is, 1 feed load for $\frac{2,400 - 30x}{30x}$ full vehicles.

$\frac{2,400 - 30x}{30x}$ + 1 = $\frac{2,400}{30x}$ vehicles, for a total cost of: $\frac{2,400}{30x}(\frac{2.57xS + 1.63xT}{5})$, for delivery at the market $\frac{2,400 - 30x}{30x}$ full loads. The cost of transport for one vehicle thus equals:

$\frac{2,400}{2,400 - 30x}(\frac{2.57xS + 1.63xT}{5})$

or

$\frac{41xS + 26xT}{80 - x}$

or, including the cost of rye:

$\frac{11,193x - 225x^2}{(182 + x)(80 - x)} + \frac{26x}{(80 - x)} = \frac{15,925x - 199.5x^2}{(182 + x)(80 - x)}$

Thünen simplifies this result to obtain : $\frac{199.5x}{182 + x}$.

C - The formula for the series of land rents as a function of distance.

a - The cost of production varies inversely and proportionally to the fertility of the soil.

    Let x = product in grain
        axT = gross product
        b" = cost of sowing
        c" = cost of preparation
        1/q = relationship between the gross product and variable costs (since variable cost only absorbs a portion of the harvest, q is a fraction)

Since $\frac{1}{q} = \frac{ax}{aqx}$, the costs proportional to the gross product are equal to aqxT.

    Let p = portion of cost expressed in money
        1-p = portion of cost expressed in grain
        hT = value of rye on the estate.

We find: the gross product $= \dfrac{ax}{h} S$

the cost of sowing $= \dfrac{b}{h} S$

the cost of preparation $= \dfrac{(1-p)c}{h} S + pcT$

The costs of cultivation and harvesting are equal then to:
$\dfrac{(1-p)aqx}{h} S + apqxT$

and the land rent is equal to:

$(\dfrac{ax}{h} - \dfrac{b + (1-p)c + (1-p)aqx}{h})S - p(aqx + c)T$

and $1S = \dfrac{hp(aqx + c)}{ax - b - (1-p)(aqx + c)} T$

But x being in the numerator and the denominator, the formula should be transformed to show the influence of x on the price.
Dividing by $(aqx + c)$:

$1S = \dfrac{hp}{\dfrac{ax-b}{aqx+c} - (1-p)}$

That is: $aqx + c = z \qquad x = \dfrac{z-c}{aq}$

From this:

$1S = \dfrac{hp}{\dfrac{az - ac - baq}{aqz} - (1-p)} = \dfrac{hp}{\dfrac{a - (\dfrac{ac+baq}{z})}{aq} - (1-p)}$

The more z rises, the more $\dfrac{ac+baq}{z}$ diminishes, and conversely. Since x and z are affected by the same variations and the total fraction diminishes when the negative portion of the denominator diminishes, the value of 1 S diminishes when x rises, and conversely.

b - Land rent, for a constant fertility, is inversely proportional to the distance.

Empirical data show that for a yield of 8S per 100 "square verges mecklembourgeoises" (obsolete measure of land area) the land rent equals : 1,168S - 641T.

Thus, for an estate located at x miles from the market, knowing that the price of an S is equal to: $\dfrac{273 - 5.5x}{182 + x} T$, the land rent equals:
$\dfrac{1,168 \,(273 - 5.5x)}{182 + x} - 641T = \dfrac{202,202 - 7,065x}{182 + x} T$.

2 - Theory of the Natural Wage.

Let : $(a+y)$ = the annual wage of a worker, a being the means of subsistence and y a surplus.

p = the product of labor or gross product less profits, insurance, and miscellaneous costs, divided by the number of workers.

Q = total capital, representing nq units of capital.

$\frac{Q}{a+y}$ thus represents the capital in terms of wages ; n workers produce np and receive n(a+y) in wages.

The surplus of the capitalist equals : $n[p - (a+y)]$ which, divided by the capital employed, gives the rate of real interest z :

$$z = \frac{n[p - (a+y)]}{np\,(a+y)} = \frac{p - (a+y)}{q\,(a+y)}$$

From which is derived : $qz(a+y) = p - (a+y)$

$$p = (a+y)(1+qz)$$

The share received by the worker is thus: $a+y = \frac{p}{1+qz}$ and that of the capitalist: $p - \frac{p}{1+qz} = \frac{pqz}{1+qz}$

They are then in a ratio :

$$\frac{\frac{p}{1+qz}}{\frac{pqz}{1+qz}} = \frac{1}{qz}$$

The gains for one year of real wages are to gains of one unit of capital as 1 is to z. Thünen believed it then possible to express in terms of labor the contribution of capital to labor. Since $\frac{Q}{a+y} = nq$, q varies as a function of the number of workers. He concluded that the entrepreneur raises by substitution the value of q so as to have the cost of the labor made by the capital and the cost of the labour made by the workers in direct proportion to their respective efficiency. But notice that z does not express the comparative efficiency of labor and capital except at the "margin of indifference". (Cf. Moore's criticism [18]).

Since the equation $a+y = \frac{p}{1+qz}$ does not permit the measurement of comparative efficiency, (a+y) being dependent on z, p on q , y and z on p, and thus p, y, and z on q, it is necessary to find p, y, and z for a given value of q.

Since, in Thünen's Isolated State, the workers have the choice between being hired or starting their own marginal farms, it appears, in the first case, that :

$$(a+y) + qz(a+y) = p, \text{ or } z = \frac{p - (a+y)}{q(a+y)},$$

equation where a, p, and q are known, y and z are unknown, and which expresses the interdependence of the wages and the interest rate.

In order to find an independant expression for the wages, Thünen assumes that a certain number of workers start a marginal farm, hiring other workers and thus dividing themselves into two groups, A and B. Group A being employed on the first farms and their surplus furnishing the means of subsistence for group B.

Let nq = the number of capitalists of group B
anq = their total yearly consumption,

then, $\frac{anq}{y}$ equals the number of capitalists in group A, since they must provide support for those in group B, and $nq + \frac{anq}{y} = \frac{nq(a+y)}{y}$ is the total number of capitalists in both groups.

Let n = the number of workers employed on the marginal farm,
n(a+y) = the total wages they are paid,
p = the product of a man working with q units of capital,
and np = the total production.

The total annual rent of the farm equals:

$$np - n(a+y) = n[p - (a+y)]$$

which is received by $\frac{nq(a+y)}{y}$ capitalists. That is, each one of them receives:

$$\frac{n[p - (a+y)]}{\frac{nq(a+y)}{y}} = \frac{y[p - (a+y)]}{q(a+y)}$$

Only y is unknown. But the workers in group B will not be satisfied by the revenue coming from their surplus times the interest rate earned on it, yz, since this is less than the rent derived by the capitalists in group B and thus by becoming entrepreneurs they could achieve a higher revenue.

It is then necessary to have: $yz = \frac{y[p - (a+y)]}{q(a+y)}$

Finally, capitalists and workers have a common interest to maximize this function. It will be maximized when its derivative with respect to y is equal to zero: that is, when $(a+y) = \sqrt{ap}$.

Thünen obtains an expression for the natural rate of interest by substituting $\sqrt{ap}$ for $(a+y)$ in the equation :

$$z = \frac{p - (a+y)}{q(a+y)} .$$

II - WEBER'S MODELS.

1 - *The Point of Minimum Transport Cost* (1).

Given a locational triangle $A_1 A_2 A_3$, let:

a_1, a_2 and a_3 = the tons of raw materials and finished products to be transported,

r_1, r_2 and r_3 = the distances expressed in miles, and P_o the unit to be located.

If P_o is situated outside the locational triangle, any displacement in the direction of one of the sides of the triangle will reduce r_1, r_2 and r_3.

If P_o is in the interior of the locational triangle the function $K = a_1 r_1 + a_2 r_2 + a_3 r_3$ is at a minimum when its derivative equals zero in all directions. This determines the point of minimum cost of transportation.

The Varignon system permits the solution of this problem mechanically. Its use is also more convenient than the geometric solution when the polygon of location involves more than three angles.

A - General case

Geometrically, if the forces a_1, a_2 and a_3 acting upon P_o are represented as straight-line segments whose length is proportional to the force (see Figure 1), each of the segments constitutes, if prolonged from P_o in its opposite direction, the diagonal of the parallelogram formed by the other two straight-line segments, provided a state of equilibrium exists.

---

(1) This model is nearly identical to that of Launhardt [15].

Figure 1

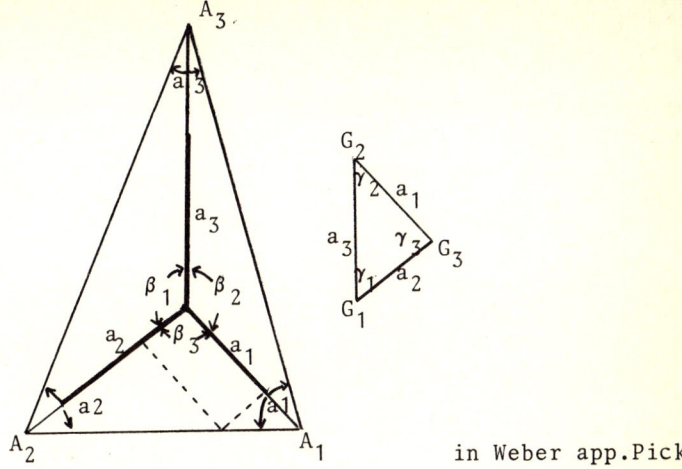

in Weber app. Pick

We obtain in this manner "the triangle of weights" $G_1G_2G_3$ whose angles γ_1, γ_2 and γ_3 are the supplements of the angles β_1, β_2, β_3 and whose sides are determined by a_1, a_2 and a_3 by virtue of the theorem of the parallelogram of forces.

P_o is determined when the position from which the lines connecting it to A_1, A_2 and A_3 form the supplementary angles of γ_1, γ_2 and γ_3 is known: that is, the position of P_o from which A_2A_3 is seen subtending the angle β_1, A_3A_1, the angle β_2 and A_1A_2, the angle β_3.

Now, the angles $\widehat{A_2P_oA_1}$ and $\widehat{A_1P_oA_3}$ will be respectively equal to β_3 and β_2 when P_o is at the intersection point of the arcs passing through A_1A_2 and A_1A_3 (see Figure 2).

Figure 2

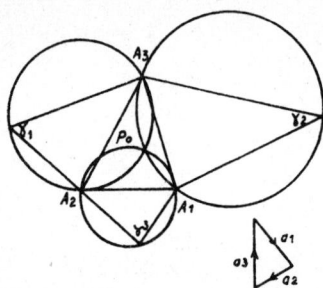

in Weber app. Pick

The arc going through A_1A_2 is obtained by constructing the angle $(\beta_3 - 90°)$ on A_1A_2 at each end. The apex C of the isosceles triangle which results gives the center of the arc. (see Figure 3). Indeed, $\widehat{A_1CA_2} = 180° - 2(\beta_3 - 90°) = 360° - 2\beta_3$; and $\widehat{C} = 2\beta_3$.

The same is true for the arc passing through A_1A_3.

Figure 3

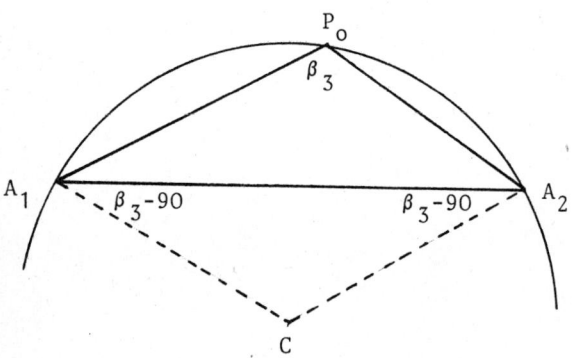

in Weber app. Pick

B - Special cases.

a - Boundary locations. If the angle β_1, for example, tends to equal 180°, at the limit there is a "line of location" and P_o finds a boundary location.

b - Corner locations. P_o will have a corner location in A_1, for example, if β_1 tends to be equal to the angle a_1 (that is, if $\gamma_1 = 180° - \hat{a}_1$).

The weight triangle $G_1G_2G_3$ can also disappear. If a_1 increases gradually, while a_2 and a_3 remain unaltered, the opposite angle γ_1 will also increase. With the increase in a_1 and of γ_1, γ_1 will equal 180° when $a_1 \geqslant a_2 + a_3$. The triangle disappears and P_o falls on A_1.

c - Case where the weights of a_1, a_2 and a_3 remain unchanged but the locational triangle changes. For example, suppose A_3 moves, while A_1 and A_2 stay fixed (Figure 4).

Figure 4

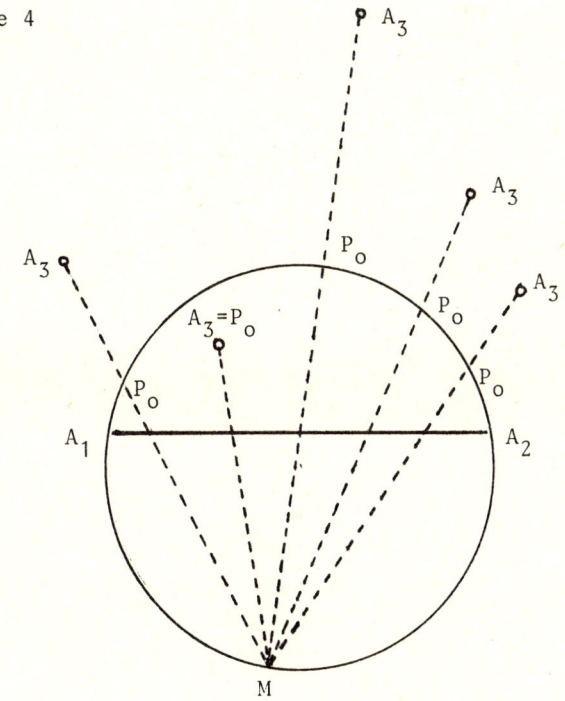

in Weber app.Pick

An arc is constructed passing through A_1A_2. If A_3 lies between the chord A_1A_2 and the arc, P_o will lie on A_3.

If A_3 lies outside the arc, P_o will be located on the arc A_1A_2. The straight line joining P_o to A_3 and those joining P_o to A_1 and A_2 form angles equal to $(180° - \gamma_2)$ and $(180° - \gamma_1)$. No matter where A_3 lies, P_o will always fall on the arc, so that the extension of A_3P_o and A_1P_o includes always the same angle γ_1 at the point P_o.

A_3P_o intersects the circle at a fixed point M.
Similarly for A_1P_o and A_2P_o.

This reasoning is correct as long as P_o does not coincide with one of the points A_1 or A_2. In these two cases, there is no longer any reason why A_3, P_o, and M should lie on one straight line. Indeed, if A_3 lies somewhere in the angular space which falls between the extension of $\overline{MA_1}$, beyond A_1 and the extension of $\overline{A_2A_1}$ beyond A_1, P_o will lie at A_1.

Similarly for A_2.

Figure 5 shows the variations of transport costs when the position of A_3 changes. One and the same position of P_o may correspond to n positions of A_3. But the total costs of transportation vary. On the figure, the positions of A_3 for which the costs of transport are equal are connected by a curve.

In the before-mentioned angular spaces beyond A_1 and A_2 the curves take the shape of concentric arcs.

In the main space lying between these two angular spaces, they take the shape of elliptic curves between the segments emerging from M and, of curves inside the circle going through A_1A_2.

Figure 5

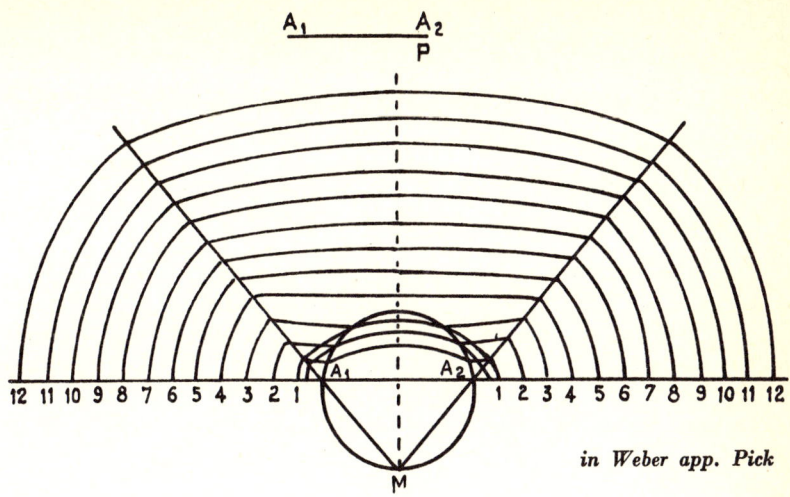

in Weber app. Pick

d - Spatial roundabout process of production. Two locational triangles $A_1A_2A_3$ and $A_1'A_2'A_3'$ are in mutual relation if A_1', for example, is at the same time the place of production of the first triangle and A_3 that for the second triangle.

Given that $P_0 = A_1'$ and $P_0' = A_3$ are the two points, we want to determine these points in such a manner that the total costs of transportation become as small as possible, given also the locations of A_1, A_2, A_3' and A_2' and the two sets of weights a_1, a_2, a_3 and a_1', a_2', a_3'.

An extreme case is where one of the two weight triangles does not exist. The problem then becomes that of determining the minimum point under the hypothesis of a locational figure with four angles.

If the two weight triangles exist, it is sufficient to apply to each of them the method developed above (see Section C). One finds, using this method, M and M' and the straight line which joins these two points cuts the two circles passing through A_1A_2 and through $A_2'A_3'$ at the required points (see Figure 6).

If the two segments subtending $\overline{A_1A_2}$ and $\overline{A_2'A_3'}$ overlap in part and MM' intersects the common part, then P_o and P_o' are combined. This is again the case of one place of production with four determining points.

If, on the other hand, MM' does not meet one of the segments at all, P_o and P_o' will lie in one of the corners.

Figure 6

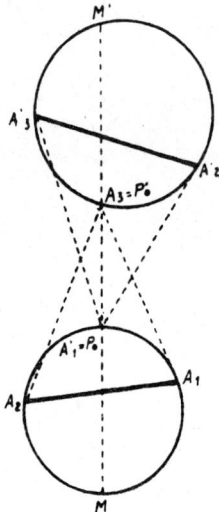

in Pick app. Weber

2 - Labor Deviation (The Isodapane Technique).

Isodapanes are geometric loci of points for which the rise in transport costs caused by the deviation from the minimum point is the same.

For very high costs of transport the isodapanes approximate circles, since the dimensions of the locational polygon become almost insignificant and it is also possible that with the increase in distance the costs of transportation might go down.

For smaller costs, the locational figure and the distribution of the set of weights exert a large influence on the shape of the isodapanes.

The transport gradient is steeper the closer the isodapanes lie to one another.

If at each of the possible positions for P_o in the plane, a perpendicular of length proportional to the sum of transport costs is raised then the set of the extreme points of these perpendiculars constitutes a surface. The lowest point of this surface lies above the point P_o; right around it the surface rises, and at a certain distance from P_o it does not differ very noticeably from the conical surface whose vertical axis goes through P_o. An object going through this surface in such a way that the distance above the plane remains constant would follow the path of one isodapane.

In the immediate neighborhood of P_o the isodapane system, and thus the shape of the transport surface, differs widely.

3 - Agglomeration.

A large unit of production having a daily production level of M will absorb (agglomerate) a small unit of production of the same kind having a daily production level of m and lying at the distance r, if the economies resulting from the agglomeration are greater than the resulting increase in transportation costs.

Assume A is the locational weight,
    Ar = the additional ton-miles for one ton of product,
    Arm = the total additional ton-miles,
    Arms = the total additional costs of transport, if s is the transport rate.

The economies which result from agglomeration are equal to $\Psi(M)$ per unit and to $M\Psi(M)$ for the daily production.

If the small unit, with a production m, is combined with the larger unit, total economies amount to:
$(M + m) \Psi (M + m)$.

Accordingly, the increase of economies due to agglomeration is equal to:
$(M + m) \Psi (M + m) - M \Psi(M)$.

Agglomeration is advantageous if
$(M + m) \Psi (M + m) - M\Psi(M) >$ Arms.

Therefore, the equation for calculating the largest distance R to which the absorbing force of the large unit of production extends is:
$$ARs = \frac{(M + m) \Psi (M + m) - M \Psi(M)}{m}$$

But the right hand side of this equation contains M as well as m. Thus, if m were at all considerable, it would indeed have an influence upon R. In fact, m is small by the nature of the problem involved, so the right hand side of the equation becomes independent of m. R becomes then a function of M alone.

Imagine that the equation contains first the value of m, and then twice the value, 2m (see Figure 7). The result is three rectangles, whose area is equal to:
$M\Psi(M)$, $(M + m) \Psi (M + m)$, $(M + 2m) \Psi (M + 2m)$,
where the largest difference:
$(M + 2m) \Psi (M + 2m) - M\Psi(M)$
approaches $(M + m) \Psi (M + m) - M\Psi(M)$ as m decreases. This gives:
$$\frac{(M + 2m) \Psi (M + 2m) - M\Psi(M)}{2m} = \frac{(M + m) \Psi (M + m) - M\Psi(M)}{m}$$
and proves the independence asserted.

Figure 7

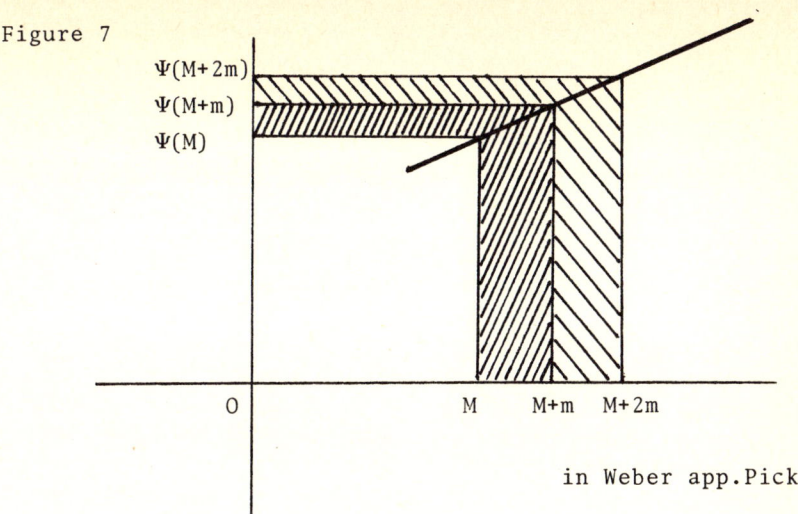

in Weber app.Pick

The function of economy or "function of agglomeration" becomes:

$$f(M) = \frac{(M + m) \Psi (M + m) - M \Psi (M)}{m}$$

The previous formula becomes: $ARs = f(M)$.

It shows that the radius within which the agglomerating force of production is effective is directly proportional to the value of the function of agglomeration and inversely proportional to the locational weight and transport rate.

The equation: $mf(M) = (M + m) \Psi (M + m) - M \Psi (M)$, represents the increments of daily economies which result when agglomeration progresses from the values M to the value M + m (see Figure 8).

Figure 8

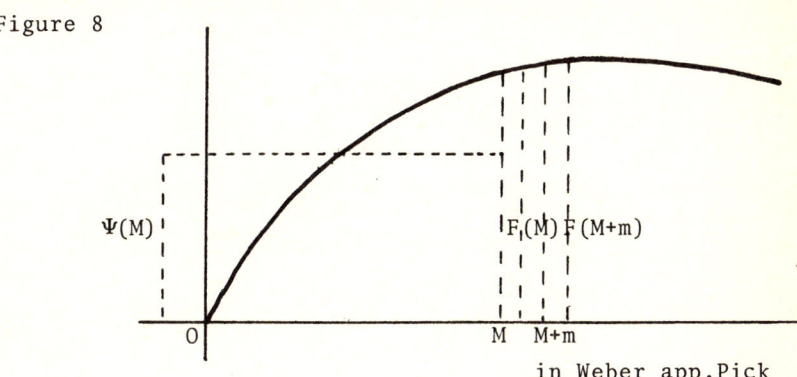

in Weber app.Pick

If a large unit of production absorbs several small ones which are distributed uniformly throughout a certain area, the radius of this area can be calculated. But bear in mind that M itself changes and increases under the influence of the process of agglomeration.

Assume φ as the amount of daily production per unit of area under the original distribution. Then, if a large unit of production G has absorbed all the small units of production just up to the circumference of the circle with radius R, it must have reached the quantum: $M = \pi R^2 \varphi$,
so that R must be: $R = \sqrt{\dfrac{M}{\pi \varphi}}$

and $\quad ARs = f(\pi R^2 \varphi)$

Or respectively $As \sqrt{\dfrac{M}{\pi \varphi}} = f(M)$.

If the value of M has been found, then it is easy to give the approximate number of large units of production which appear in the area. For if Ω indicates the amount of daily production in the entire area, the number will be: Ω / M.

The problem is then to determine M in such a way that:

$$\dfrac{As}{\sqrt{\pi \varphi}} \cdot \sqrt{M} = f(M) .$$

The diagram of the function of agglomeration permits the determination of the size of agglomeration.

Imagine that a second curve is drawn representing f(M) by coordinating each abscissa M to the ordinate:

$$N = \dfrac{As}{\sqrt{\pi \varphi}} \cdot \sqrt{M} \quad \text{(see Figure 9)}$$

A parabola results. The required abscissa is one for which the curve of f(M) and the parabola have equal ordinates.

Figure 9

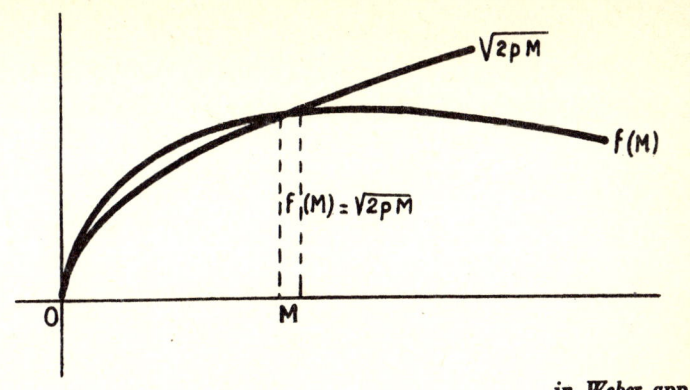

in Weber app. Pick

There exist several possibilities. The curve f(M) may extend below the parabola and remain below it. In that case the equation is never fulfilled and N is always greater than f(M), which means that the increments in transport costs are always larger than the economies of agglomeration.

Second, the curve f(M) may run above the parabola and then cross it at some point, remaining below it after that. In this case, agglomeration will occur up to the point of intersection of the two curves.

The third case, in which f(M) always lies above the parabola does not, it seems, correspond to any actual cases.

III - PALANDER'S MODELS (1).

1 - *The Delineation of the Market*.

Given two points, A and B, separated by a distance, 1, competing for their surrounding market, let P_A and P_B be their respective f.o.b. prices and F_A, F_B, the freight rate per ton-mile for their respective products (Figure 10).

Figure 10

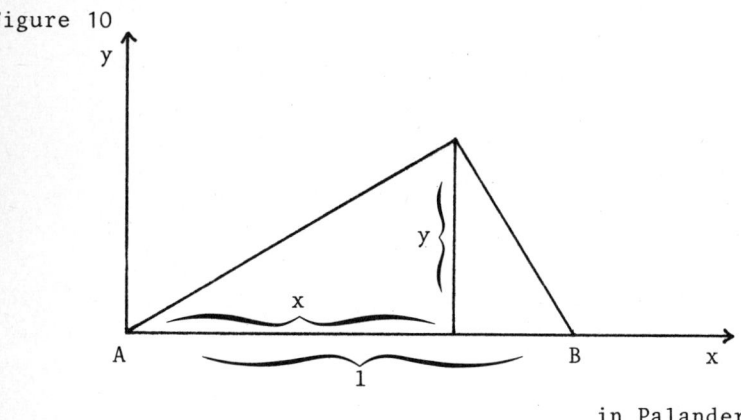

in Palander

If x and y are the coordinates of the points of indifference of the buyers, there are to be obtained at these points using the Pythagorean theorem:

$$P_A + F_A \sqrt{x^2 + y^2} = P_B + F_B \sqrt{(1 - x)^2 + y^2} .$$

This gives the equation of the isostante.

A - If $P_A = P_B$ and $F_A = F_B$, the equation becomes:

$$\sqrt{x^2 + y^2} = \sqrt{(1 - x)^2 + y^2}$$

---

(1) Some of the results presented here differ from those obtained from the same basic equations by Palander; this is no doubt due to some errors. For a comparison of the results see Ponsard (C.) [244, 83-114].

Or : $x^2 + y^2 = 1^2 + x^2 - 2lx + y^2$

and simplifying:

$$x = \frac{1}{2}$$

The isostante is then the perpendicular bisector of AB.

B - If $P_A = P_B$, the equation becomes:

$$F_A \sqrt{x^2 + y^2} = F_B \sqrt{(1 - x)^2 + y^2}$$

If it is assumed that: $\frac{F_A}{F_B} = K > 1$,

then : $K \sqrt{x^2 + y^2} = \sqrt{(1 - x)^2 + y^2}$

which becomes, raising it to the square and factoring out:

$$x^2(K^2 - 1) + 2lx + y^2(K^2 - 1) = 1^2$$

or, dividing by $(K^2 - 1)$:

$$x^2 + \frac{2lx}{K^2 - 1} + y^2 = \frac{1^2}{K^2 - 1}$$

That is : $(x + \frac{1}{K^2 - 1})^2 + y^2 = \frac{1^2}{K^2 - 1} (1 + \frac{1}{K^2 - 1})$

and : $(x + \frac{1}{K^2 - 1})^2 + y^2 = \frac{K^2 1^2}{(K^2 - 1)^2}$

The isostante is then a circle of radius $\frac{Kl}{K^2 - 1}$, whose center is at a distance $\frac{1}{K^2 - 1}$ from point A, and B is inside the circel. (Palander's demonstration includes the case studied by Schneider (E.) [88].

C - If $F_A = F_B = F$, the equation becomes:

$$P_A + F \sqrt{x^2 + y^2} = P_B + F \sqrt{(1 - x)^2 + y^2}$$

If $P_B > P_A$, assume $w = \frac{P_B - P_A}{F}$ and obtain:

$$\sqrt{x^2 + y^2} - w = \sqrt{(1 - x)^2 + y^2}$$

Squaring and simplifying, the equation becomes:

$$- 2w \sqrt{x^2 + y^2} = 1^2 - w^2 - 2\mathrm{l}x$$

Squaring once again:

$$w^2 (x^2 + y^2) = \frac{(1^2 - w^2)^2}{4} + 1^2 x^2 - 1x(1^2 - w^2)$$

After rearranging terms, the equation becomes:

$$x^2(1^2 - w^2) - 1x(1^2 - w^2) - w^2 y^2 = -\frac{(1^2 - w^2)^2}{4}$$

or, dividing by $-\frac{(1^2 - w^2)^2}{4}$, gives:

$$(1^2 - w^2)(x - \tfrac{1}{2})^2 - w^2 y^2 = -\frac{(1^2 - w^2)^2}{4} + \frac{1^2}{4}(1^2 - w^2)$$

or :

$$\frac{(x - \tfrac{1}{2})^2}{w^2} - \frac{y^2}{(1^2 - w^2)} = -\frac{1}{4} [(1^2 - w^2)^2 - (1^2 - w^2)1^2]$$

Dividing by $w^2(1^2 - w^2)$ and by $1/4$ and simplifying:

$$\frac{(x - \tfrac{1}{2})^2}{\frac{w^2}{4}} - \frac{y^2}{\frac{1^2 - w^2}{4}} = 1$$

In this case the isostante is a hyperbola whose center $(1/2, 0)$ cuts AB at a distance $\frac{1 + w}{2}$ from A. (Palander's demonstration includes the case studied by Fetter (F.A.) [44]).

D - If $1 = 0$, that is, if A and B fall at the same point, the equation becomes:

$$P_A + F_A \sqrt{x^2 + y^2} = P_B + F_B \sqrt{x^2 + y^2}$$

or : $(F_A - F_B) \sqrt{x^2 + y^2} = P_B - P_A$

or : $x^2 + y^2 = \left(\dfrac{P_B - P_A}{F_A - F_B}\right)^2$

The isostante is then a circle of center (A,B) and radius $\dfrac{P_B - P_A}{F_A - F_B}$.

E - If $y = 0$, that is, if the market is reduced to the straight line AB, the equation becomes:

$$P_A + F_A \sqrt{x^2} = P_B + F_B \sqrt{(1 - x)^2}$$

or : $P_A + F_A |x| = P_B + F_B |1 - x|$

The isostante is then reduced to one point whose position depends on the relative prices and the relative costs of transportation.

2 - Price Policy and Transport Costs.

Assume a linear market of length MN = 1 on which are uniformly distributed a set of buyers whose demand is inelastic and for which two sellers, A and B, are competing.

Let MA = a, BN = b;

K_A and K_B = the unit costs of production of the sellers;
P_A and P_B = their f.o.b. prices;
G_A and G_B = their unit profits;
F_A and F_B = their unit transport cost;
Q_A and Q_B = their outputs;
E_A and E_B = their total profits;
x and y = the distances from the places of production to the limit P of their markets (see Figure 11).

Figure 11

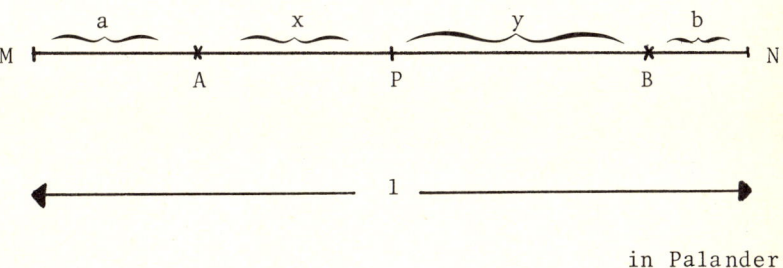

in Palander

At P : $K_A + G_A + F_A x = K_B + G_B + F_B y$

and since $x = 1 - a - y - b$ and $y = 1 - a - x - b$:

$$x = \frac{K_B - K_A + G_B - G_A + F_B (1 - a - b)}{F_A + F_B}$$

and $y = \dfrac{K_A - K_B + G_A - G_B + F_A (1 - a - b)}{F_A + F_B}$

But, $Q_A = a + x$ and $Q_B = b + y$

Thus simplifying:

$$E_A = Q_A G_A = \left(\frac{K_B - K_A + G_B - G_A + F_B (1 - b) + F_A a}{F_A + F_B}\right) G_A$$

and $E_B = Q_B G_B = \left(\dfrac{K_A - K_B + G_A - G_B + F_A (1 - a) + F_B b}{F_A + F_B}\right) G_B$

The profits of each seller depend on the profits of the other. In the case of duopoly of double dependency, each assumes the unit profit of his competitor to be constant. The total profits will be maximized when the partial derivatives of E_A and E_B with respect to G_A and G_B are simultaneously equal to zero, that is:

$$\frac{\delta E_A}{\delta G_A} = \left(\frac{K_B - K_A + G_B + F_B (1 - b) + F_A a}{F_A + F_B}\right) - \frac{2G_A}{F_A + F_B} = 0$$

and $\dfrac{\delta E_B}{\delta G_B} = \left(\dfrac{K_A - K_B + G_A + F_A (1 - a) + F_B b}{F_A + F_B}\right) - \dfrac{2G_B}{F_A + F_B} = 0$

Multiplying the first equation by 2 and combining it with the second:

$K_B - K_A + 2F_B 1 - F_B b + F_A a + F_A 1 = 3G_A$

Thus:

$G_A = 1/3 (K_B - K_A) + 1/3 (2F_B + F_A)1 + 1/3 (F_A a - F_B b)$

and $G_B = 1/3 (K_A - K_B) + 1/3 (2F_A + F_B)1 - 1/3 (F_A a - F_B b)$

Total profits are equal to:

$$E_A = \frac{(K_B - K_A + 2F_B 1 + F_A 1 + F_A a - F_B b)^2}{9 (F_A + F_B)}$$

and $E_B = \dfrac{(K_A - K_B + 2F_A l + F_B l - F_A a + F_B b)^2}{9 (F_A + F_B)}$

Palander's demonstration includes on the one hand Laundhardt's theory, which assumes a = b = 0, and gives:

$$G_A = 1/3 (K_B - K_A) + 1/3 (2F_B + F_A) l$$

and $E_A = \dfrac{[K_B - K_A + 1(2F_B + F_A)]^2}{9 (F_A + F_B)}$.

A similar procedure gives G_B and E_B [16]. It also includes, on the other hand, Hotelling's theory, which assumes $K_B = K_A = 0$ and $F_B = F_A = F$, and gives:

$$G_A = F(1 + \tfrac{a-b}{3}), \quad G_B = F(1 - \tfrac{a-b}{3}), \quad E_A = \tfrac{F}{2}(1 + \tfrac{a-b}{3})^2$$

and $E_B = \tfrac{F}{2}(1 - \tfrac{a-b}{3})^2$ [65] .

By assuming zero cost of production, it is possible to introduce prices, which are identified by unit profits.

If the transport rates are the same:

$$E_A = Q_A G_A = Q_A P_A = (a+x) P_A$$

or: $E_A = P_A \left(\dfrac{P_B - P_A + F(1 - a - b)}{2F} + a \right)$

or: $E_A = \tfrac{1}{2} [P_A(1 + a - b) - \dfrac{P_A^2}{F} + \dfrac{P_A \cdot P_B}{F}]$

and $E_B = \tfrac{1}{2} [P_B(1 - a + b) - \dfrac{P_B^2}{F} + \dfrac{P_A \cdot P_B}{F}]$

A given profit results from various combinations of P_A and P_B. The profit curves are hyperbolas, whose vertices are turned downward and whose asymptotes have as their respective equations:

$P_A = 0$ and $P_B = P_A - F(1 + a - b)$, that is I'I"

$P_B = 0$ and $P_A = P_b - F(1 - a + b)$, that is H'H"

(see Figure 12 where the prices are given along the axis. The values given to the symbols are those of Hotelling, that is l=35, a=4, b=1, and F=1; the values for E_A and E_B are shown in the figure).

Figure 12

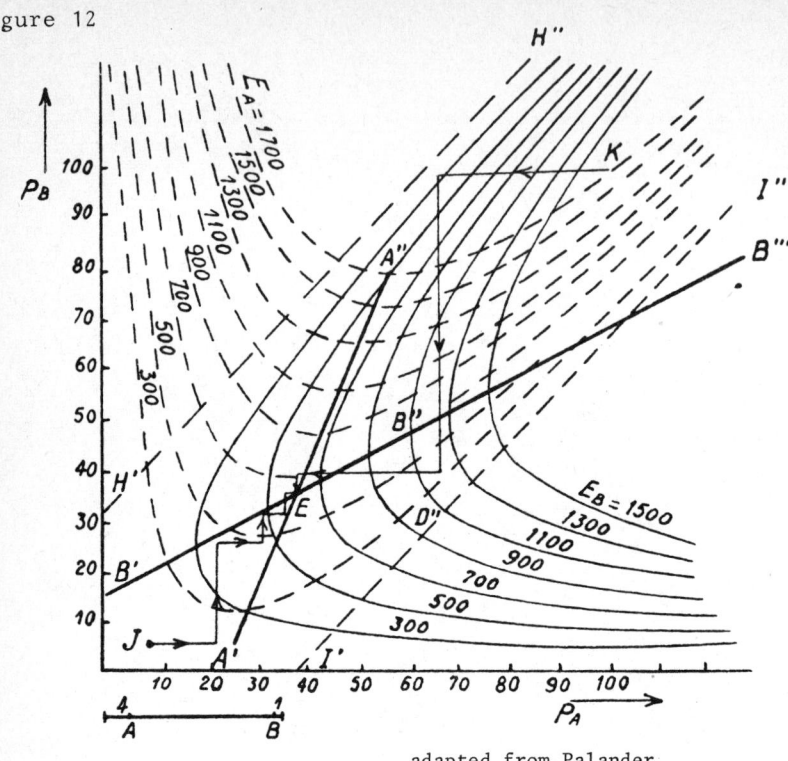

adapted from Palander

Every variation of P_A or P_B is expressed by a horizontal or vertical displacement, each seller tending towards the higher profit curve determined by the price of his competitor, that is, towards the tangent to the line expressing this price.

In determining the points that have a horizontal or vertical tangent, for all the profit curves, the geometric locus of all the price combinations that result from the adjustment of the prices between them are obtained.

Given A'A" for A and B'B" for B, the equations are obtained by equating to zero the derivatives of the preceding profit functions with respect to P_A and P_B respectively. Thus:

$$P_A = \frac{P_B}{2} + \frac{F(1 + a - b)}{2} \quad \text{(for A'A")}$$

and $$P_B = \frac{P_A}{2} + \frac{F(1 - a + b)}{2} \quad \text{(for B'B")}$$

Thus price adjustments will occur until the point E is reached, at the intersection of A'A" and B'B", and at this point a stable equilibrium will be attained.

However, figure 11 shows that if $P_A < P_B - F(1 - a - b)$, B will no longer be able to sell, and if $P_A > P_B + F(1 - a - b)$, A will be expelled from the market. Thus, the line C'C", whose equation is $P_A = P_B - F(1 - a - b)$, and the line D'D", whose equation is $P_A = P_B + F(1 - a - b)$, determine the limits of the field of competition (see Figure 13). Beyond them the lines A'A" and B'B" do not have any economic significance.

Figure 13

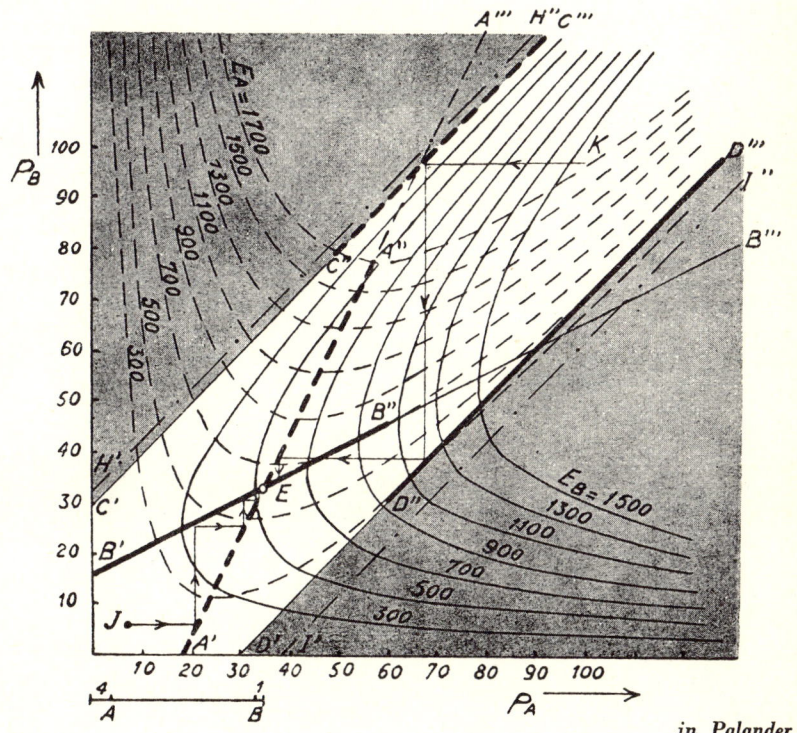

in Palander

Even within these limits, it is possible that a portion of the adjustment lines will have no value for the changes in prices.

The seller A, for example, has in effect a choice between two policies: (1) to fix his price according to A'A" or (2) to fix it according to C'C".

In the first case, since : $P_A = \dfrac{P_B + F(1+a-b)}{2}$
there results from the profit equation:

$$E_A = \dfrac{[P_B + F(1 + a - b)]^2}{8F}$$

In the second case, since $P_A = P_B - F(1-a-b)$, there results, using the same procedure:

$$E'_A = [P_B - F(1 - a - b)](1 - b)$$

Thus, if P_B is low, the first policy is more profitable; if P_B is high, the second one is more profitable. The profit of A is equal in both cases for the value $P_B = F(31 - 3b - a)$, which can be obtained by setting $E_A = E'_A$. A will choose the first policy when P_B is lower than this value and the second one when it is higher. (see Figure 13: for $P_B < 78.3$, A follows A'A" and for $P_B > 78.3$, he follows C'C").

Similarly, B adjusts his price following B'B" or D'D" according to whether: $P_A \gtreqless F(31 - 3a - b)$.

If A'A" and B'B" intersect at a point where they both determine the prices a stable equilibrium will arise. Given the equations of A'A" and B'B", all depends on the values of a, b, and 1. In figures 12 and 13, the equilibrium conditions are met. This is not the case for figure 14, for example, (where a=4, b=15 and 1=35) where the price suffers a continuous oscillation (C'"A"B"C'").

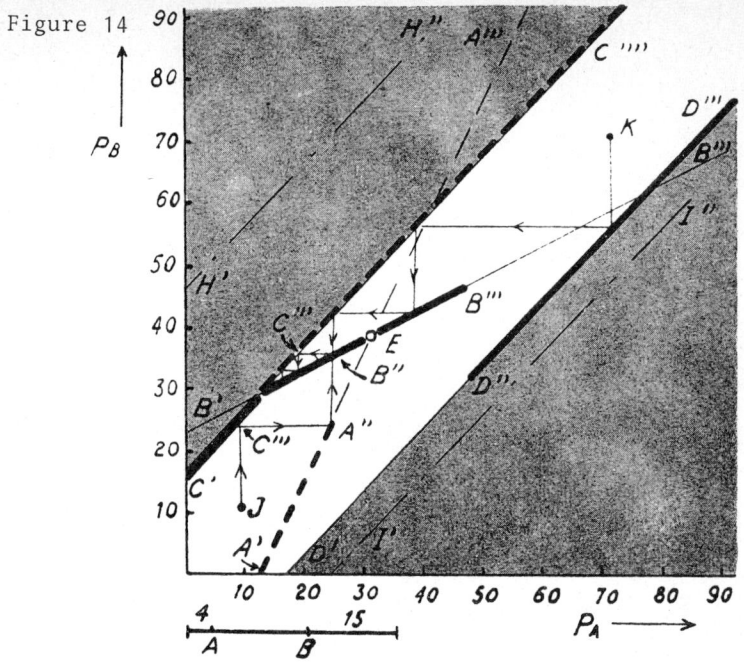

Figure 14

in Palander

In order that a stable equilibrium be reached, it is necessary to have the lines A'A" and B'B" intersect at a point where both of them still serve effectively as lines of adjustments.

In the equilibrium state, Hotelling's equation gives:

$$P_B = F(1 - \frac{a-b}{3}),$$

A goes from sharing the market to the elimination of B when $P_B = F(31 - 3b - a)$. Solving this system of equations with respect to b:

$$b = \frac{31 - a}{5}$$

A similar reasoning is valid for a. Finally, the equilibrium conditions are met when:

$$a \leqslant \frac{31 - b}{5} \quad \text{and} \quad b \leqslant \frac{31 - a}{5}$$

Figure 15 shows that equilibrium is only possible for the combinations of values a and b located above and to the left of the lines where equations respond to the competitors conditions (white surface): that is, if the competitors have small hinterlands.

Figure 15

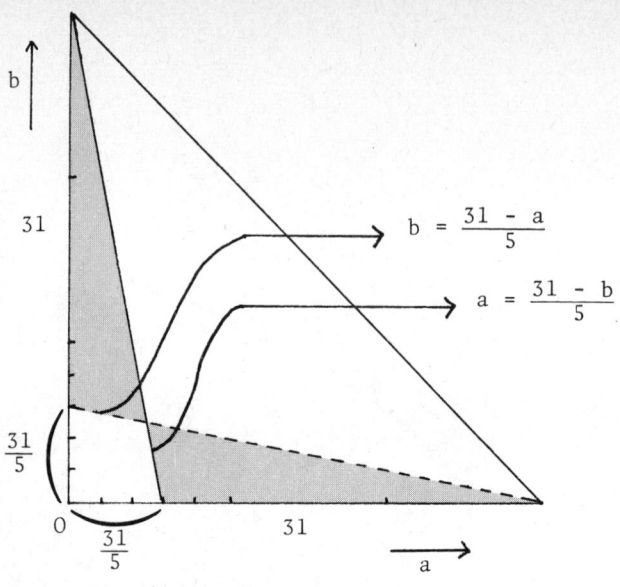

3 - *Isoline Technique.*

a - Definitions.

Isodistances : loci of points for which the straight line distances are equal.

Isochrones : loci of points for which the durations of transportation are equal.

Isotims : loci of points for which the prices of a given product are equal

Isovectors : loci of points for which the costs of transportation for a given product are equal.

Isostantes : loci of points for which the difference in prices for 2 or n products equals the difference in costs of transport.

Isodapanes : loci of points for which the sum of transport costs of several products are equal.

b - Construction of isodapanes.

Isodapanes cannot be obtained directly, as can the other isolines. First, it is necessary to draw isovectors. For the case of two points on a transport surface, the isodapanes go through all the intersection points of the isovectors for which the sum of transport costs is the same (see Figure 16. In this figure the isodapane for which the transport costs equal 15 has been drawn). For three points, first isodapanes are constructed between two points and then the isodapanes that go through the intersection points of isovectors of the third point with the isodapanes of the first two are added. The technique is similar for more than three points.

Figure 16

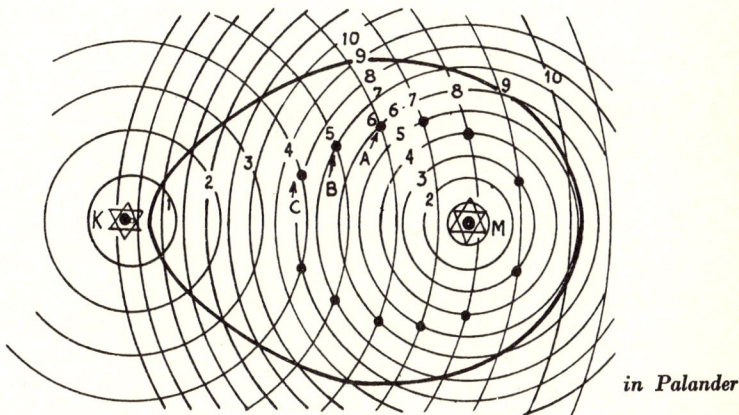

in Palander

c - Isodapanes and the point of minimum transportation cost.

Under the conditions described above, the point of minimum transportation cost is included inside the lowest isodapane.

Given the simple assumption that the raw materials are obtained at one point, A, and the product is delivered at another point, B, Figures 17 and 18 then show that if there is a uniform freight rate a neutral zone appears between A and B, and if there is a variable freight rate two minimum zones appear around A and B.

Figure 17   Figure 18

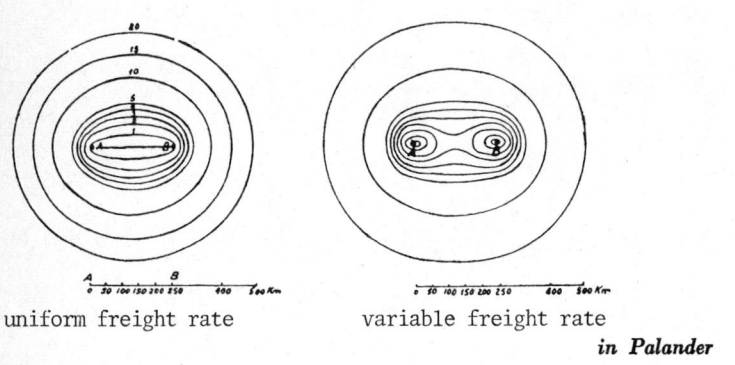

uniform freight rate   variable freight rate

in **Palander**

Under more complex assumptions, isodapanes vary according to the observed data and give the desired solution. For example, if the raw material weighs twice as much as the finished product, Figures 19 and 20 show that whether there is a uniform or variable freight rate, the point of minimum transportation cost is found at the raw material source. In the case of variable freight rate, at a distance which is about the same for each of two locations, a zone arises where the costs of transportation are nearly constant.

The conclusions are similar in the case of n transportations.

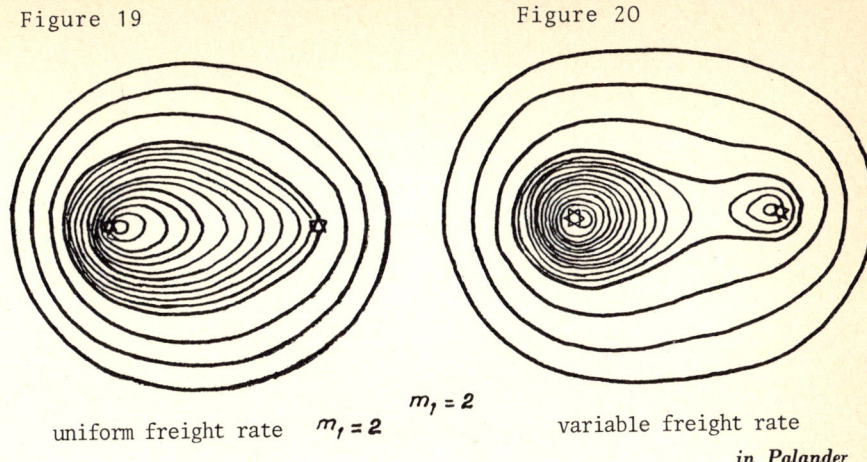

Figure 19   Figure 20

uniform freight rate $m_1 = 2$ $m_1 = 2$ variable freight rate

in Palander

d - Isodapanes and deviations from the minimum point.

According to the data in the analysis, the location of production can be displaced within certain limits. Figures 21 to 24 show, for different location polygons, that it is possible to displace the production throughout the whole figure without the transport cost going up more than 20% and that within certain limits the transportation factor is no longer the most essential factor of location.

Figures

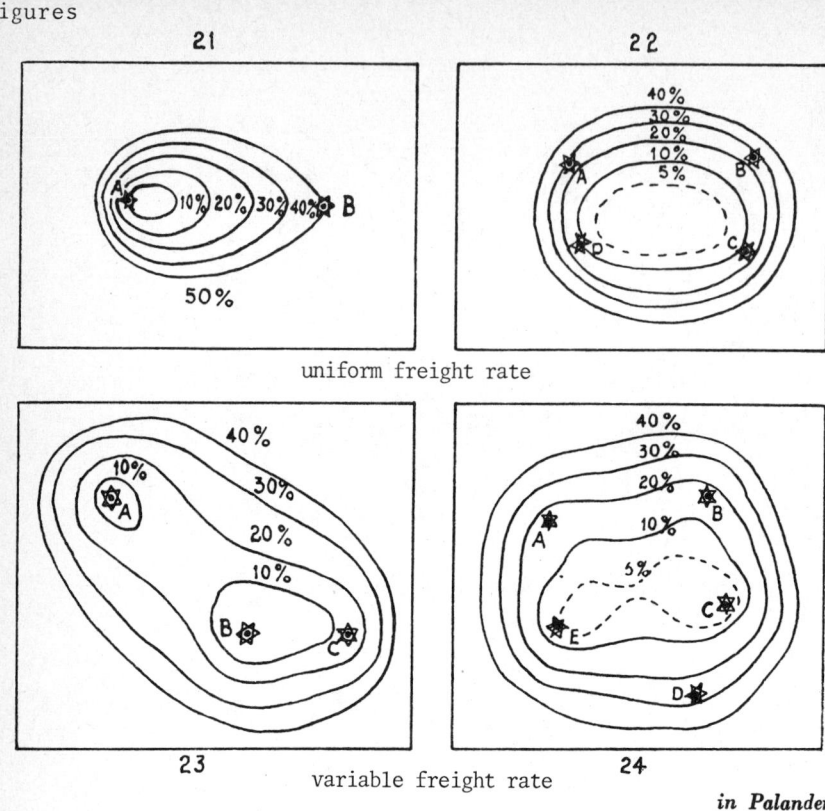

uniform freight rate

variable freight rate

in Palander

There might also be the problem of choice among several point sources of raw materials. Isodapanes are constructed in such a way that they connect the points for which the sum of the purchase price and total transport cost is equal for all sources. The definitive isodapanes then are determined by the intersection of several elementary systems of isodapanes. Figure 25 illustrates the simplest assumption of the choice between M and M', assuming that the weight of the raw material is double that of the product and that the price at M is lower by two monetary units to that at M'. Two systems of isodapanes result and the area where M' is preferred is separated from that where M is preferred by a hyperbola. When the prices are equal, this hyperbola becomes a straight line going through K, the place of consumption.

Figure 25

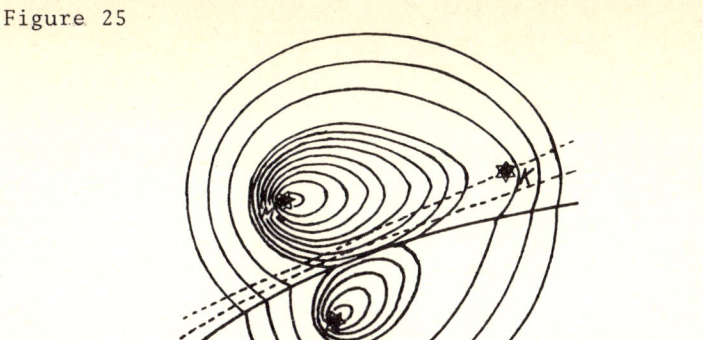

uniform freight rate *in Palander*

m = 2

e - Isodapanes and agglometations of locations.

Obstacles that prevent product from being transported in a straight line cause modifications of the isovectors and thus of the isodapanes and locations.

Figure 26 illustrates the case of a transport surface cut by a river, with a limited number of bridges. The isodapanes take the form $AC_1G_1G_2$, $AC_2G_2G_3$, etc. Thus, an agglomeration of several independent locations is born at C_3.

Figure 26

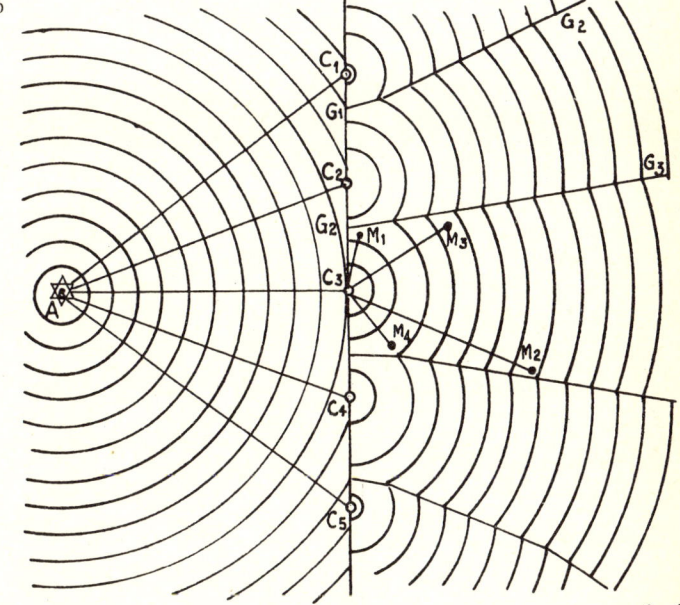

in Palander

f - Transport lines and the minimum point.

Finally, consider the irregular configuration of transport lines and any one sequence of sources of raw materials, places of consumption, and junction points. Figure 27, constructed under the assumption of a uniform freight rate, shows that the point of minimum transport cost is B, but that the cost of the transport is almost as low at A, C and M_2 .

Figure 27

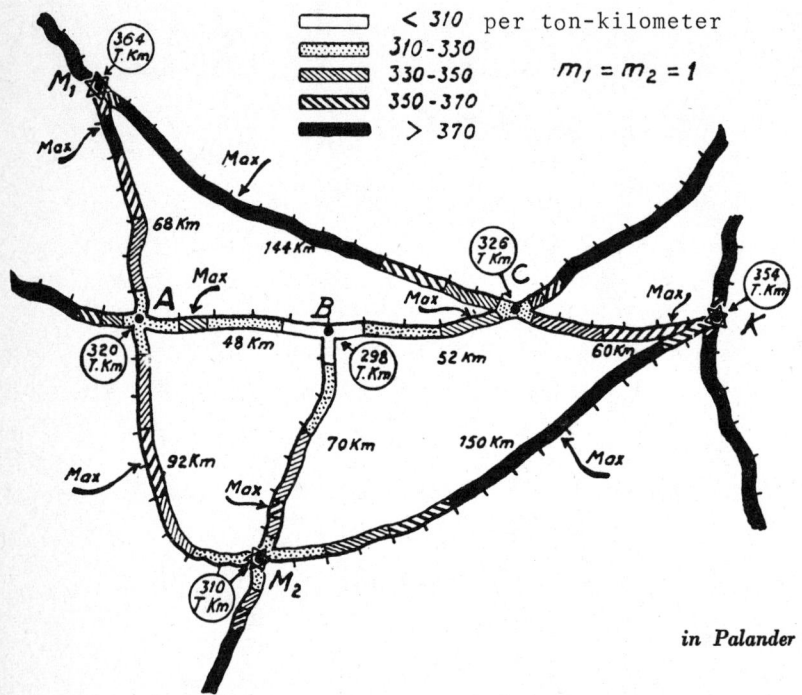

in Palander

4 - Law of Refraction.

A - Given a combined transportation from a point A to a point B separated by a line DE which delimits two transport surfaces on which the costs of transportation differ (see Figure 28), the problem becomes that of determining a point C, located on DE such that the course ACB brings about minimum transportation costs.

Figure 28

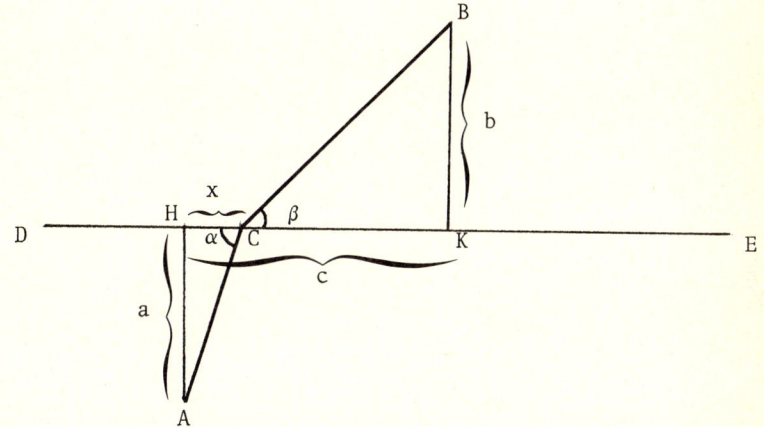

in Palander

If the transportation costs per ton-kilometer are equal to KF (K > 1) on the surface A and to F on the surface B, the equation of the transport expenditures T is given by:

$$T = KF \sqrt{a^2 + x^2} + F \sqrt{(c-x)^2 + b^2}$$

And T will be a minimum for:

$$\frac{dT}{dx} = \frac{KFx}{\sqrt{a^2 + x^2}} - \frac{F(c-x)}{\sqrt{(c-x)^2 + b^2}} = 0$$

but $\quad \dfrac{x}{\sqrt{a^2 + x^2}} = \dfrac{CH}{CA} = \cos \alpha$

and $\dfrac{c-x}{\sqrt{(c-x)^2 + b^2}} = \dfrac{CK}{CB} = \cos \beta$

Thus, in order for T to be a minimum:

$$\cos \alpha = \dfrac{\cos \beta}{K}$$

When K=1 (case of equality of transport cost on the two surfaces), the transportation route is a straight line. The more K rises, the more AC approaches the perpendicular to the line DE.

B - The assumption of combined transportation from a point A on a transport surface to a point B on a transport line approaches the preceding case when the surface B is reduced to the line DE. Then β becomes zero and the transport costs are a minimum for $\cos \alpha = 1/K$. Thus, the larger K, the smaller $\cos \alpha$ and the more AC approaches the perpendicular to the transportation line DE.

C - If transportation takes place from A, on one side of the transportation line DE to B', on the other side, or to B" on the same side as A, the value of K determines the slope of the lines AC', B'C" and B"C''' with respect to DE. In this case the angles AC'D, B'C"E and B"C''' E are equal to α (see Figure 29).

Figure 29

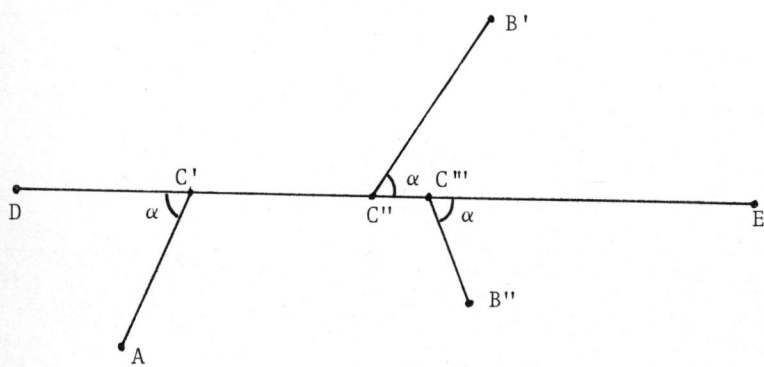

in Palander

5 - Conditions of Competition and Location.

A - Assymetric duopoly.

If B behaves "superpolitically", that is, aware of his competitor's reaction path to his own choice of location and price, he adjusts his price according to the adjustment line A'A" whose equation is:

$$P_B = 2P_A - F(1 + a - b)$$

and he will do so in such a way that, once A has adapted his own price to that of B, P_A will correspond to that level of A'A" located on B's highest profit curve (see Figure 30).

Figure 30

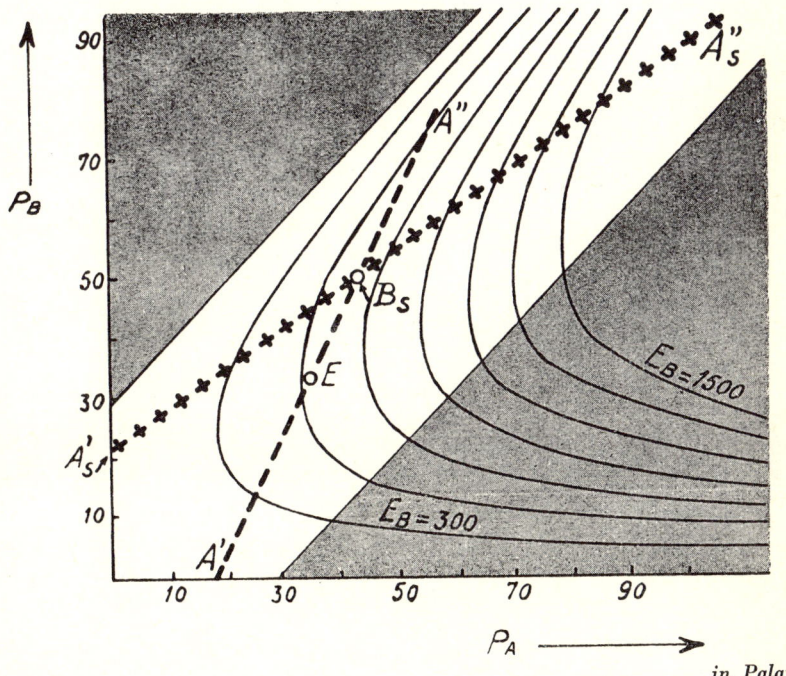

in Palander

E_B rises along A'A" up to a certain point and then starts diminishing; thus E_B will be a maximum when A adjusts his price at that point where A'A" is tangent to the profit curve.

The derivative of the profit curve can be written as:

$$\frac{\frac{d E_B}{d P_A}}{\frac{d E_B}{d P_B}} = \frac{d P_B}{d P_A}$$

and since: $E_B = 1/2 [P_B (1-a+b) - \frac{P_B^2}{F} + \frac{P_A \cdot P_B}{F}]$

then: $\frac{d E_B}{d P_A} = \frac{P_B}{2F}$

and: $\frac{d E_B}{d P_B} = \frac{F(1-a+b) - 2 P_B + P_A}{2F}$

Thus: $\frac{d P_B}{d P_A} = \frac{P_B/2}{1/2 \, F(1-a+b) - P_B + P_A/2}$

For A'A", whose equation can be written as:

$2P_A = P_B + F(1+a-b)$

it is true that: $\frac{d P_B}{d P_A} = 2$

From these two equations can be deduced:

$P_B = 2 F(1-a+b) - 2 P_B + 2 P_A$

or $P_B = \frac{2}{3} F(1-a+b) + \frac{2}{3} P_A$

which is the equation of the geometric locus A'sA"s of the points in the profit curves of B which have a tangent parallel to A'A".

The intersection point Bs of A'sA"s and A'A" has as its coordinates, obtained by multiplying the equation of A'A" by -1 and subtracting it from the equation of A'sA"s:

$\frac{F}{3} (5l+a-b) - \frac{4}{3} P_A = 0$

Then: $P_A = \frac{F}{4} (5l+a-b)$

and, substituting this value in the equation of A'A":

$P_B = 2 [\frac{F}{4} (5l+a-b)] - F(1+a-b)$

or : $P_B = \frac{F}{2}(31-a+b)$

Therefore, B picks a price equal to: $\frac{31-a+b}{2}$, and A, a price: $\frac{51+a-b}{4}$

But if A prefers to lower his price, following the line C'C", and attempts to conquer the whole market, the equilibrium is broken. Since A goes from sharing the market to attack or defense when $P_B = F(31-3b-a)$, the line A'A" is only valid up to $P_B = \frac{1}{2}(31-a+b)$ as long as $b \leq \frac{31-a}{7}$. If b goes beyond this value, either B fixes the maximum price compatible with A'A", up to a certain value for b, or, for any value higher than the value set for b, B fixes the highest price for which A will not go from defense to attack. The value of b for which the transition takes place can be deduced from the values of P_B for which A ceases to follow D'D" in order to follow C'C". This value is: $b = \frac{(1-a)^2}{1-2a}$.

B - Duopoly of double control.

If both sellers behave "superpolitically", A in turn adjusts his price following the line B'B" and sets a price P_A such that at its level, P_B will be fixed at a point corresponding to that level of B'B" located on A's highest profit curve.

The equation of B'B" being $P_B = \frac{P_A + F(1-a+b)}{2}$

Then: $\frac{d P_A}{d P_B} = 2$

A's profit equation being:

$$E_A = \frac{1}{2}[P_A(1+a-b) - \frac{P_A^2}{F} + \frac{P_A \cdot P_B}{F}]$$

Then: $\frac{d E_A}{d P_B} = \frac{P_A / 2}{F(1+a-b) - 2P_A + P_B}$

Thus: $2 = \frac{P_A}{F(1+a-b) - P_A + P_B}$

and $P_B = \frac{3}{2} P_A - F(1+a-b)$

which is the equation of the locus of the points at which A's profit curves have parallel tangents to B'B": that is, B'sB"s (see Figure 31, where B'sB"s cuts B'B" at Bs).

Figure 31

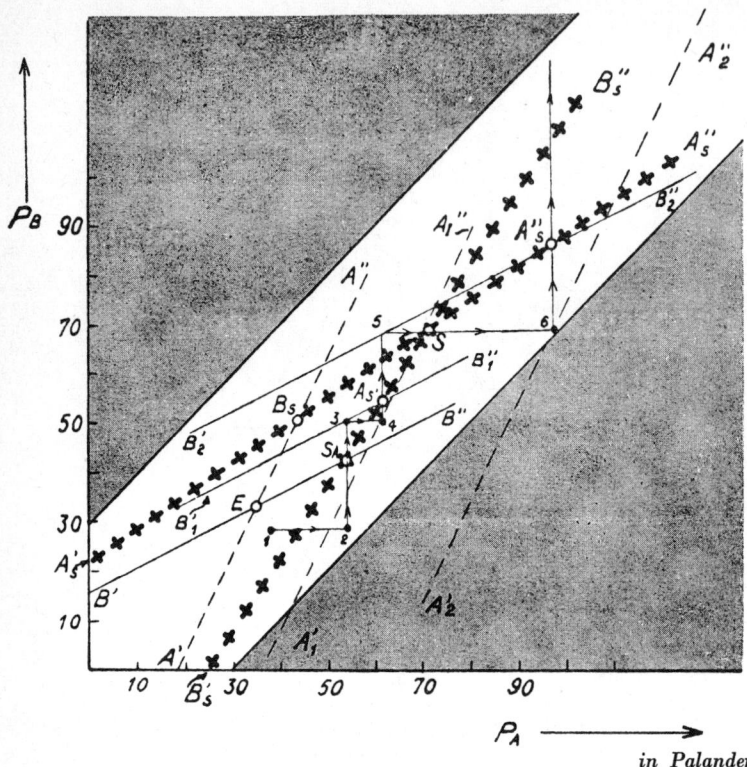

in Palander

At a given price 1, A, beleiving that B behaves according to B'B", chooses a price corresponding to point 2 and expects that B will raise his price up to As. The result will be price 3. Meanwhile, B expects that A will establish his price at Bs, and A expects that B will establish his price at As.

An equilibrium state is not possible as long as either one of the sellers does not modify his conception of the other's policy. If they modify instead their estimation of the costs, A may think that B behaves, for example, according to $B'_1B"_1$ and will establish price 4; B, on the other hand, may think that A behaves, say, according to $A_1'A_1''$, and he will establish his price at 5. But then A may believe that B behaves according to $B_2'B_2''$ and establishes a price 6; B will think that A behaves according to $A_2'A_2''$, and so on. The prices do not show any tendency towards equilibrium; they rise continuously so that the assumptions made a priori are no longer valid.

C - The methods of fixing prices.

a - If the seller B calculates his prices from some other starting point, F, than the point of production, B, and if b and c are constant (see Figure 32), the equations for A'A" and B'B" respectively become, for the case of duopoly of double dependence:

$$P_B = 2 P_A - F(1+a-b)$$

and $\quad P_B = \dfrac{P_A}{2} + \dfrac{F(1-a+2b-c)}{2}$

and at their intersection, the equilibrium point:

$$P_A = F(1 + \frac{a-c}{3})$$

and $\quad P_B = F(1 - \dfrac{a+c}{3} + b - c)$

Figure 32

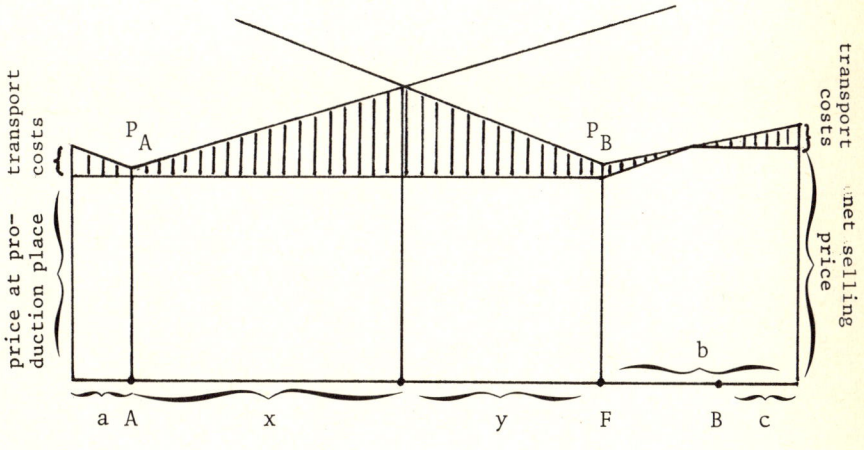

in Palander

Thus, A's price is identical to that obtained under the assumption that the sellers compute their prices from the place of production, while B's price is higher. The farther the point F is from B, the higher will be the equilibrium prices.
A's profit remains equal to: $E_A = \frac{F}{2} (1 + \frac{a-c}{3})^2$
while for B the profit becomes: $E_B = \frac{F}{2} (1 - \frac{a-c}{3})^2 + F(b^2 - c^2)$
the rise in profit due to the rise in price between F and B and from B to the limit of the market on B's side.

If the position of F is variable, F will tend to approach A, B's profit rising with (b-c).

This same reasoning can be applied to the case where A or B, or A and B behave "superpolitically".

b - If B practices uniform delivered pricing, while A fixes his price at the place of production (see Figure 33), there results, for the case of duopoly of double dependence, that at those points at which the consumer buys indifferently from one or the other seller:

$P_A + F x = P_B$

so: $x = \dfrac{P_B - P_A}{F}$

$y = 1 - a - b - \dfrac{P_B - P_A}{F}$

Since: $Q_A = a + x$ and $Q_B = b + y$

then: $E_A = a P_A + \dfrac{P_A (P_B - P_A)}{F}$

and $E_B = P_B (1-a) - \dfrac{P_B^2 - P_A \cdot P_B}{F} - \dfrac{F b^2}{2} - \dfrac{F (1-a-b - \frac{P_B - P_A}{F})^2}{2}$

Figure 33

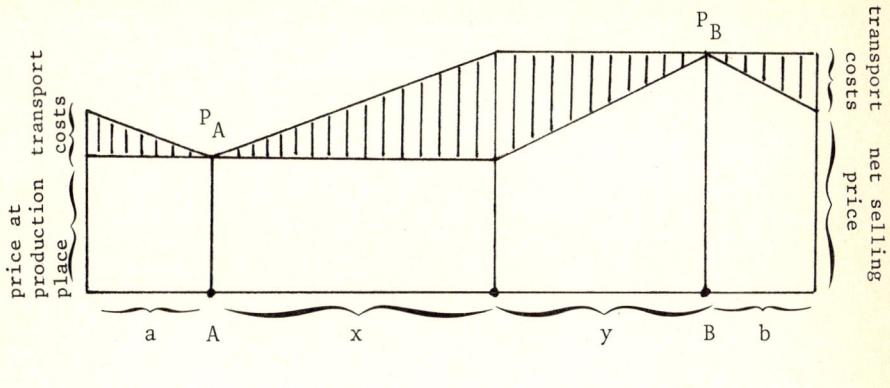

in Palander

Taking the derivative of E_A with respect to P_A and E_B with respect to P_B we get the equations of the adjustment lines of A and B:

$$P_B = 2 P_A - aF$$

$$P_B = \frac{2}{3} P_A + \frac{F}{3} (21 - 2a - b).$$

At the equilibrium point:

$$P_A = \frac{F(21 + a - b)}{4} \quad \text{and} \quad P_B = \frac{F(21 - a - b)}{4}$$

Thus: $Q_A = \frac{21 + a - b}{4}$

and: $Q_B = \frac{21 - a - b}{4}$

$$E_A = \frac{F(21 + ab)^2}{16}$$

and $E_B = \frac{3(21-a)^2 + 6b(21-a) - 29b^2}{32}$

The profits are lower than those obtained under the assumption of pricing from the place of production.

Nevertheless B is interested in practicing uniform delivered pricing as long as A has not reacted following his adjustment line. In fact, if $l = 35$, $a = b = 5$, and $F = 1$, Hotelling's equations give:

$$P_A = P_B = 35 \text{ and } E_A = E_B = 612.5$$

while if B practises a uniform delivered pricing, Palander's equations, before the reactions following the adjustment lines, give:

$$P_B = 41.7 \text{ and } E_B = 624$$

But when the new equilibrium is reached:

$$P_B = 30 \text{ and } E_B = 312.5$$

The advantage of changing the pricing method disappears.

This kind of reasoning can also be applied to the case where one of the sellers behaves "superpolitically".

D - The choice of location.

If a is given (see Figure 11), while b is variable, B's choice of location will depend, in the case of duopoly of double dependence, on his strategy.

a - If he attempts to conquer the whole market:

$$P_B = P_A - F(1 - a - b)$$

and $E_B = 1 \, [P_A - F(1 - a - b)]$

E_B varies directly with b and B will tend to approach A as much as possible.

b - If B defends his hinterland:

$$P_B = P_A + F(1 - a - b)$$

so: $E_B = b \, [P_A + F(1 - a - b)]$ which will be maximized if:

$$\frac{\delta E_B}{\delta b} = P_A + F(1-a) - 2Fb = 0$$

Thus: $b = \dfrac{P_A + F(1-a)}{2F}$

The location of B depends on A's price. If $P_A \geq F(1-a)$, b must be equal to or greater than $(1-a)$, so that E_B is a maximum. This condition being impossible, B will again fix b as large as possible. If $P_A < F(1-a)$, B will tend to locate farther from A the lower P_A.

c - If B wants to share the market:

$$P_B = \frac{P_A}{2} + \frac{F(1 - a + b)}{2}$$

and: $E_B = \dfrac{[P_A + F(1 - a + b)]^2}{8F}$

Thus B tends to approach A.

IV - LÖSCH'S MODELS.

1 - *Location of two agricultural products.*

The independent variables of the model that explain the location of two agricultural products, I and II, simultaneously cultivated, are:
- A : the outlay, per hectare, in marks
- E : the yield per hectare, expressed in 1/2 quintal (per centner)
- p : the market price per 1/2 quintal
- k : distance from the market in kilometers
- f : freight cost per 1/2 quintal per kilometer

The dependant variables of the model are:
- $a = \frac{A}{E}$: the outlay per 1/2 quintal, in marks
- $\pi = p - kf$: local price per 1/2 quintal
- $r = \pi - a$: local profit per 1/2 quintal
- $R = E(p - kf) - A$: rent per hectare
- $m = p - a$: highest profit per 1/2 quintal

Product I will be cultivated at the center of a given area and product II at the periphery, when $R_1 > R_2$ at the center, or:

$$E_1(p_1 - kf) - A_1 > E_2(p_2 - kf) - A_2$$

But, at the center of the area, $k = 0$, so:

$$E_1 p_1 - A_1 > E_2 p_2 - A_2 .$$

Thus, the first condition is:

$$1 < \frac{E_1 p_1 - A_1}{E_2 p_2 - A_2} \qquad (1)$$

Product II will be cultivated at the periphery under the condition $R_1 < R_2$ at the periphery, or:

$$E_1 (p_1 - kf) - A_1 < E_2 (p_2 - kf) - A_2 .$$

But $kf = p_2 - a_2$, since the maximum profit per 1/2 quintal fixes the largest margin of tolerable transport costs.

From this:

$$E_1 p_1 - E_2 p_2 + E_1 a_2 - A_1 < E_2 p_2 - E_2 p_2 + E_2 a_2 - A_2$$

or, by replacing a_2 by A_2/E_2, multiplying both sides by E_2/E_1 and simplifying:

$$E_1 p_1 - E_1 p_2 + \frac{E_1 A_2}{E_2} - A_1 < 0$$

$$E_1 p_1 - A_1 < E_1 p_2 - \frac{E_1 A_2}{E_2}$$

$$E_1 p_1 - A_1 < \frac{E_1}{E_2} (E_2 p_2 - A_2)$$

Dividing by $(E_2 p_2 - A_2)$, the second condition is obtained:

$$\frac{E_1 p_1 - A_1}{E_2 p_2 - A_2} < \frac{E_1}{E_2} \qquad (2)$$

Combining both conditions:

$$1 < \frac{E_1 p_1 - A_1}{E_2 p_2 - A_2} < \frac{E_1}{E_2} \qquad (3)$$

These inequalities express the conditions that must be met by A, E, p so that the cultivations of products I and II are concentric.

Table 1 shows 27 cases in which these magnitudes vary, using all possible relative magnitudes. Condition (1) is always fulfilled in 7 cases, so that only product I is cultivated. Condition (2) is always fulfilled in 7 other cases, product II being cultivated everywhere. In 12 additionnal cases, the assumptions about the relative magnitudes of A, E and p are not sufficient to decide between the different crops, so that it is not known whether one, or both, will be cultivated. The two products will both be planted when the two preceding conditions are fulfilled or when neither is met. This depends on the numerical values of the variables. Nevertheless, cases 10 and 18 differ from the other ten. Since in both cases $E_1/E_2 = 1$, both conditions can never be met. A similar effect results if: $(p_1-p_2) = (a_1-a_2)$. In effect:

$(p_1-p_2) = (a_1-a_2) = \frac{A_1}{E_1} - \frac{A_2}{E_2}$; but since $E_1 = E_2$, it follows that:

$(p_1-p_2) = \frac{A_1-A_2}{E_1} = \frac{A_1-A_2}{E_2}$; and from that: $E_1(p_1-p_2) = A_1-A_2$ or $E_1p_1-A_1 = E_2p_2-A_2$ and thus: $\frac{E_1p_1 - A_1}{E_2p_2 - A_2} = 1$, an expression which is incompatible with conditions (1) and (2). The two products are then grown side by side in adjacent sectors.

It is only in the remaining ten cases, where certain relative values for A, E, and p exist in which both conditions are fulfilled, that the choice of a crop is a function of distance and that Thünen's type rings appear.

Product I will be grown in the interior as long as $E_1 > E_2$. Condition (3) can be written, replacing A for Ea, as:

$$1 < \frac{E_1(p_1-a_1)}{E_2(p_2-a_2)} < \frac{E_1}{E_2}$$

or: $E_2(p_2-a_2) < E_1(p_1-a_1) < E_1(p_2-a_2)$

or: $E_2m_2 < E_1m_1 < E_1m_2$. Since $m_1 < m_2$, it follows that $E_1 > E_2$. Thus, at the market place, product I brings in the greater total profit E_1m_1 and the smaller profit per unit m_1.

As distance grows, transport costs absorb m_1 up to a limit where the loss is no longer counterbalanced by a superior yield per hectare. At this limit $R_1 = R_2$ and cultivation of crop II starts. Then:

Table 1

case N°	relative size of variables			variables fulfill condition N°						crop produced					
				always		sometimes		never		monoculture		polyculture			
				1	2	1	2	1	2	unconditioned	conditioned	unconditioned	conditional		
												cultivated together	rings		
	E_1 to E_2	A_1 to A_2	p_1 to p_2							I	II	I or II	I and II side by side	I inside	II inside
1		>	>	.	.	x	x	.	.	.	.	o	.	o	.
2			=	.	.	x	x	.	.	.	.	o	.	o	.
3			<	.	.	x	x	.	.	.	.	o	.	o	.
4	>	=	>	x	.	.	.	.	x	o	.	.	.	.	.
5			=	x	.	.	.	.	x	o	.	.	.	.	.
6			<	.	.	x	x	.	.	.	.	o	.	o	.
7		<	>	x	.	.	.	.	x	o	.	.	.	.	.
8			=	x	.	.	.	.	x	o	.	.	.	.	.
9			<	.	.	x	x	.	.	.	.	o	.	o	.
10		>	>	.	.	x	x	.	.	.	.	o	.	o	.
11			=	.	x	.	.	x	.	.	o	.	.	.	.
12			<	.	x	.	.	x	.	.	o	.	.	.	.
13	=	=	>	x	.	.	.	.	x	o	.	.	.	.	.
14			=	.	.	.	.	x	x	.	.	.	o	.	.
15			<	.	x	.	.	x	.	.	o	.	.	.	.
16		<	>	x	.	.	.	.	x	o	.	.	.	.	.
17			=	x	.	.	.	.	x	o	.	.	.	.	.
18			<	.	.	x	x	.	.	.	.	o	.	o	.
19		>	>	.	.	x	x	.	.	.	.	o	.	.	o
20			=	.	x	.	.	x	.	.	o	.	.	.	.
21			<	.	x	.	.	x	.	.	o	.	.	.	.
22	<	=	>	.	.	x	x	.	.	.	.	o	.	.	o
23			=	.	x	.	.	x	.	.	o	.	.	.	.
24			<	.	x	.	.	x	.	.	o	.	.	.	.
25		<	>	.	.	x	x	.	.	.	.	o	.	.	o
26			=	.	.	x	x	.	.	.	.	o	.	.	o
27			<	.	.	x	x	.	.	.	.	o	.	.	o

in Lösch

$E_1(p_1-kf) - A_1 = E_2(p_2-kf) - A_2$. From that:

$$k = \frac{E_2 p_2 - E_1 p_1 + A_1 - A_2}{f(E_2-E_1)}$$ and since $m = p-a$ and $a = A/E$:

$$k = \frac{1}{f}\left(\frac{E_1 m_1 - E_2 m_2}{E_1 - E_2}\right)$$

Thus the distance k of the line of transition from one crop to the other is inversely proportional to the freight costs and to the difference in the yields and directly proportional to the difference in the total profits. This can be written:

$$kf = \frac{E_1 m_1 - E_2 m_2}{E_1 - E_2}$$

This equation expresses the condition of equality between the marginal transport cost and the marginal revenue per 1/2 quintal. The model solves two problems simultaneously: the volume of production and its location.

Finally, the Thünen's type rings can be inverted. If the harvest of product I (potatoes) from one hectare, expressed in weight, is equal to double that of product II (corn), it is known that the rings will be established if:

$$m_1 < m_2 \quad \text{and} \quad \frac{E_1}{E_2} > \frac{m_2}{m_1}$$

And since $E_1 = 2E_2$, the conditions can be written: $m_1 < m_2 < 2m_1$. Besides, since $a_2 = 2a_1$: $p_1-a_1 < p_2-2a_1 < 2(p_1-a_1)$.

Geometrically, if distance is plotted on the abscissa and prices on the ordinate, the curve of maximum profit per 1/2 quintal of product I cuts the price axis farther from (and the distance axis nearer to) the origin than the curve of product II, since product I gives the highest total profit, but the smallest profit per unit (see Figure 34 (a)). The maximum possible distance from the market is determined by the point of intersection with the distance axis.

If $p_1 = 6$ marks and $p_2 = 12$ marks per 1/2 quintal, the demand for product I will have to be equal to double that for product II and the cultivated surface will be the same for the two products. Assume a greater elasticity of demand for product I so that the two curves have the same slope on Figure 34(b) and (d). If the market takes a linear form, the rings then become rectangles and the supply is proportional to their breadth (see Figure 34(c)).

Figure 34

Figure 35

in Lösch

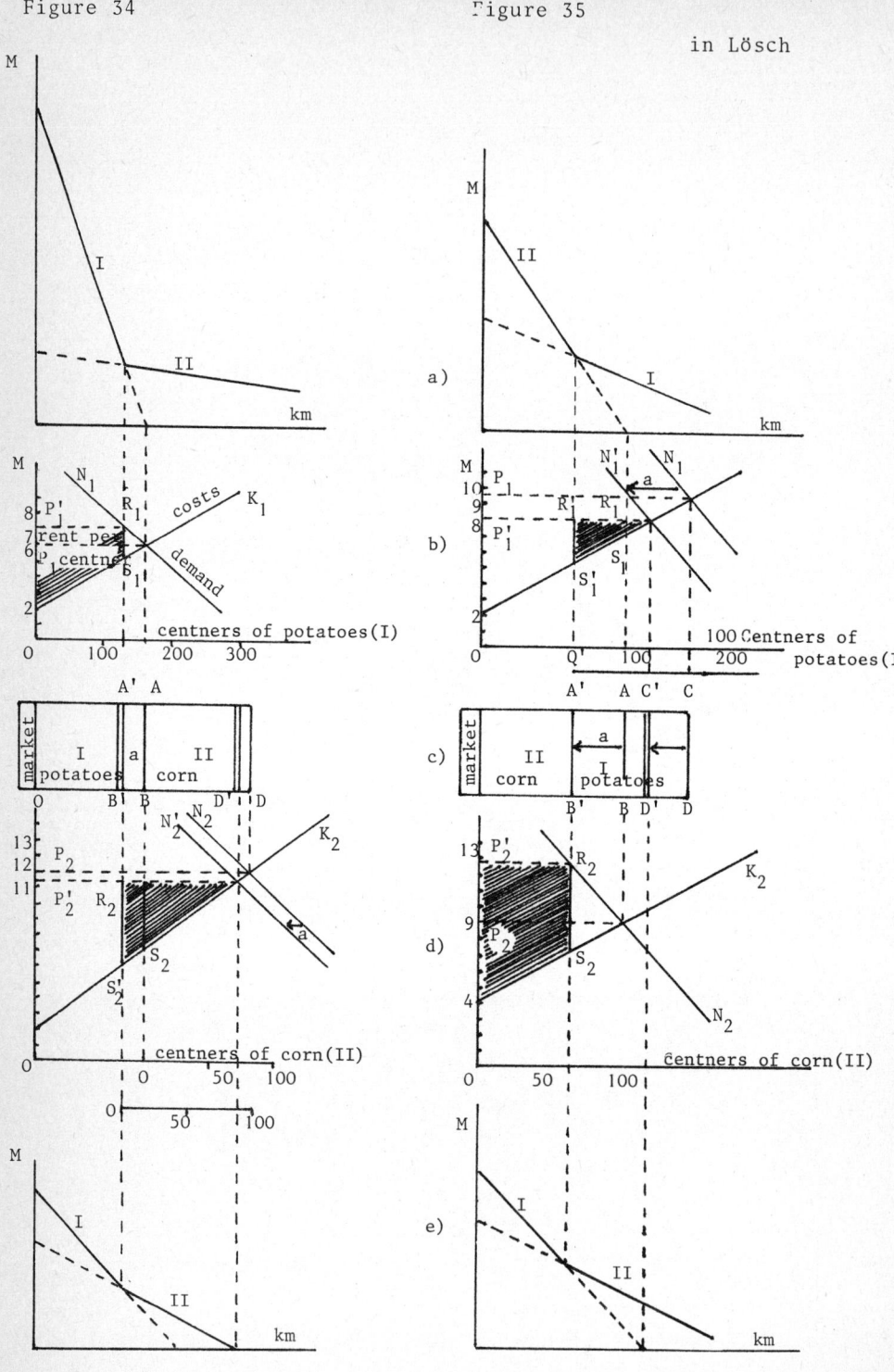

The production of each product is considered to be in equilibrium if neither product yields a rent at its respective margin of cultivation. But at the boundary line AB (on Figure 34(c)), if the rent of the product I is zero, that of product II, on the other hand, is equal to R_2S_2. The cultivation of product II is extended at the expense of product I, price p_2 falls to p_2' and price p_1 rises to p_1'. The cultivation of product I is then diminished by the area AA'BB' and total cultivation is diminished by the smaller area CC'DD'. A general equilibrium situation is thus obtained when: $R_1S_1 = \frac{R_2'S_2'}{2}$. That is, when the profit per 1/2 quintal of product I equals one half of the profit per 1/2 quintal of product II.

To obtain the inversion of the rings, it is sufficient to make only one modification in the preceding assumptions. It is sufficient to assume that product I is being imported and that the fields lying near the market are not available for cultivation of product I. Cultivation of product I will then be assigned to the periphery. But an equilibrium will only appear when the cultivation of product I is extended beyond the boundary AB on Figure 35(c) up to the boundary A'B! The profit per 1/2 quintal of product II is then equal to double that of product I and the rents per hectare are the same.

Moreover, this equilibrium is relatively stable. If on Figure 36 is drawn the rent curve (dotted line) which will result in the inversion, the rent curve after the inversion (continuous line), and the rent curve when no inversion occurs (broken line), a mean rent zone will appear (hatched area between E and F), where the farmers are not interested in a reversal of locations. Only a few farmers, located between O and E and between F and H, are interested in reversal. The double transition to the cultivation of product I between O and E, and to the cultivation of product II, between F and H, will lower the respective prices of both and will diminish the rent between E and F.

Figure 36

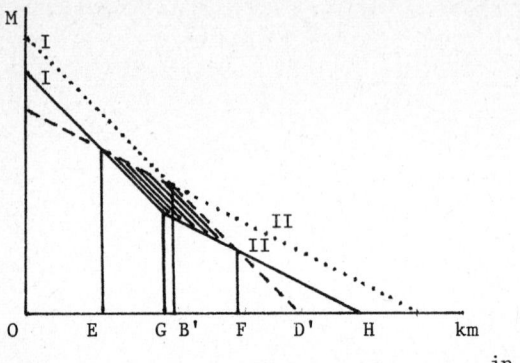

in Lösch

2 - The Equations of General Spatial Equilibrium.

A series of coordinates (x,y) is assigned to each product to designate the position of its different places of production on a surface which is assumed to be uniform and plane.

Thus $p_1^1(x_1^1 y_1^1)$ designates the location of production of good 1 at place 1; similarly $p_2^1(x_2^1 y_2^1)$ designates the location of production of good 1 at place 2; and generally $p_q^m(x_q^m y_q^m)$ designates the location of production of good m at place q.

Similarly, there is one equation describing the boundary of the market area for each production site.

Thus, $(\alpha_1^1, \beta_1^1, \ldots, \epsilon_1^1)$ is the abbreviation for the equation of the market boundary of good 1 produced at site 1. Similarly, $(\alpha_1^2, \beta_1^2, \ldots, \eta_1^2)$ is the abbreviation for the equation of the market boundary of good 2 produced at site 1; and generally, $(\alpha_q^m, \beta_q^m, \ldots, \epsilon_q^m)$ abbreviates the equation of the market boundary of the good m produced at site q.

Table 2 summarizes the mathematical symbols for the equations of the spatial arrangement.

Table 2

Product N°	Place of Production		Market Boundaries	
	Site	Number	Abbreviations of their equations	Number
1	$P_1^1\ (X_1^1 Y_1^1);\ P_2^1\ \ldots\ P_a^1$	a	$\alpha_1^1, \beta_1^1\ \ldots\ \epsilon_1^1;\ \alpha_2^1, \beta_2^1\ \ldots$	A
2	$P_1^2\ (X_1^2 Y_1^2);\ P_2^2\ \ldots\ P_b^2$	b	$\alpha_1^2, \beta_1^2\ \ldots\ \eta_1^2;\ \alpha_2^2, \beta_2^2\ \ldots$	B
...	...	...	...	...
m	$P_1^m (X_1^m Y_1^m);\ \ldots\ P_q^m$	q	$\alpha_1^m, \beta_1^m\ \ldots\ \delta_1^m;\ \alpha_2^m, \beta_2^m\ \ldots$	Q
m	(together)	n	(together)	N
		$= \Sigma\ a+b+\ldots+q$		$= \dfrac{\Sigma\ A+B+\ldots+Q}{2}$

in Lösch

The given variables of the model are the following:

$d^m = f^m(\pi)$ = individual demand for product m

$\pi_q^m = \Phi^m(D_q)$ = f.o.b. price of product m at site q

$k_q^m = \chi^m(D_q)$ = average cost of production of product

$D_q^m = \psi(f^m,\ x_q^m y_q^m,\ \alpha_q^m \beta_q^m\ \ldots\ \epsilon_q^m,\ \sigma,\ \sigma_q^m\ \ldots)$ = total demand for m at q

$S_q^m = D_q^m \cdot (\pi_q^m - k_q^m)$ = the profit on product m at site q

σ = the rural population per square kilometer

σ_q^m = the urban population of the town P_q^m

r = the freight rate

m = the number of products

G = the size of the entire area.

The unknowns of the model are the following:

		Number of Unknowns
π_q^m =	factory price of product m at location P_q^m	n
G_q^m =	sales area of location P_q^m in square kilometers	n
q^m =	number of towns that produce product m	m

	Number of Unknowns
x_q^m, y_q^m = coordinates of location P_q^m	$2n$
α_q^m, β_q^m, ... ϵ_q^m = equations for the boundaries of the market area of P_q^m	N
Sum:	$4n + m + N$

A series of equations satisfying the five conditions of equilibrium can then be proposed. In Table 3 these conditions are restated on the left side while the corresponding equations appear on the right. It is easy to verify that the number of equations equals the number of unknowns.

Table 3

	Condition	Equation which fulfills condition	Number of Equations
Maximum number of Producers {	1. Maximum Profit	$\dfrac{\delta S_q^m}{\delta X_q^m} = 0;\ \dfrac{\delta S_q^m}{\delta Y_q^m} = 0$	$2n$
	2. Total area used	$\Sigma G_1^m + \Sigma G_2^m + ... + G_q^m = G$	m
	3. No unusual profits	$\Phi^m(D_q) = \chi^m(D_q)$	n
	4. Area as small as possible	$\dfrac{\delta \pi_q^m}{\delta G_q^m} = \dfrac{\delta k_q^m}{\delta G_q^m}$	n
	5. Boundaries as Indifference lines	$\alpha_q^m = \pi_q^m + r_q^m \sqrt{(X-X_q^m)^2 + (Y-Y_q^m)^2}$ $= \pi_{q-1}^m + r_{q-1}^m \sqrt{(X-X_{q-1}^m)^2 + (Y-Y_{q-1}^m)^2}$	N
		SUM :	$4n + m + N$

1. Equation of maximum individual profit: if all the locations are given, except one, then all the parameters of the equation of S_q^m are constant except for the coordinates (x_q^m, y_q^m). When the partial derivates are zero with respect to those coordinates then the point p_q^m is optimum.
2. The sum of the individual areas equals the entire area unless it is assumed that the latter is larger than necessary for m firms but not large enough to take $(m+1)$.
3. Unusual profits are zero when the f.o.b. prices equal the average costs of production.
4. An area is as small as possible when the demand curve is tangent to the costs curve. Equation 3 expresses the first condition of tangency and equation 4 the second.
5. For any given point (x,y) at the boundary of a sales area, the c.i.f. price (f.o.b. price + transport costs x distance) paid by the buyer must be the same whether he buys at p_q^m or at p_{q-1}^m.

3 - *Economic Regions.*

Imagine an area which is homogeneous and plane containing self-sufficient and equidistant farms and a population continuously distributed. If any of the farms produce a manufactured good over and above its needs in order to sell it, its market will be extended over the neighboring area whose dimensions will depend on the demand curve and delivered prices (see Figure 37).

Figure 37

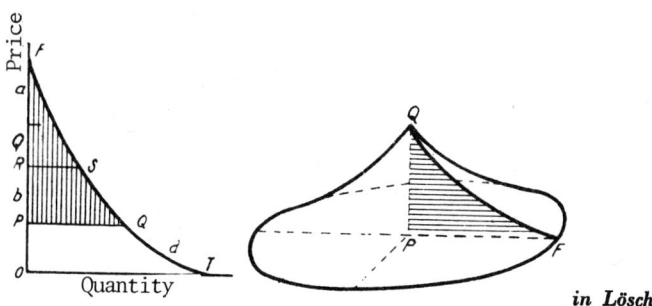

in Lösch

Given:
d = f(p+t) = individual demand
D = total demand as a function of the f.o.b. price p
p = OP = f.o.b. price
PQ = sales
t = shipping costs per unit
R = PF = the greatest possible shipping costs
b = twice the population of a square in which it costs 1 mark to ship one unit along one side

In Figure 37 the total volume of sales equals the volume of the cone generated by pivoting PQF around PQ.

The volume of this cone is equal by definition to: $2\pi y_0 \cdot F$.

Since: $y_0 \cdot F = \int_0^R f(p+t) \cdot t \cdot dt$

it becomes: $2\pi \int_o^R f(p+t).t.dt$

Finally, if the population density b/2 is taken into account, the total demand is equal to:

$D = b. \pi \int_o^R f(p+t).t.dt$

The price p being in turn a function of the total demand, the good can only be produced if the cost curve is lower or at most equal to that of the demand. When this last condition is met, the quantity supplied is determined and the radius of the sales circle is equal to the radius of the demand surface that can be served at the prevailing price.

Figure 38 shows how a network of hexagons replaces the circles as the ideal shape for the markets when the sellers are numerous and their sales radii contiguous. The area of the hexagon is slightly smaller than the area of the circle which circumscribes this hexagon, so that the individual demand curve tends to be displaced towards the left.

Figure 38

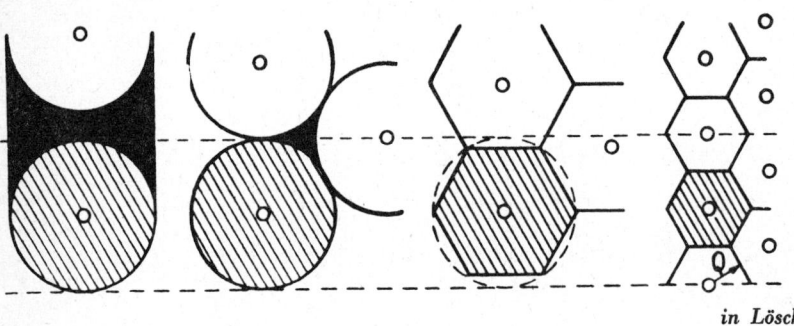

in Lösch

If the assumption is introduced that population is distributed in a discontinuous fashion and concentrated on several farms, which are still assumed to be equidistant from one another, Figures 39(a), (b) and (c) show that according to the inclination of the hexagons, the same center of production (B_1 for example) can supply a variable number of farms. There typical orientations exhaust the possibilities available.

(1) Fig.39(a). The farms are located at the angles of the hexagons. The distance nV between the seller $A_1(=B_1)$ and the farthest buyer $A_2, A_3, \ldots, A_7$ is equal to a (the distance between A_1 and A_2, A_2 and A_3, etc.). Due to the competition between B_2 and B_3 at A_2, of B_3 and B_4 at A_3, etc., B_1 does not entirely supply 7 centers, but the equivalent of 3: that is, all of A_1 plus 1/3 of the other 6 farms $A_2, A_3, \ldots, A_7$. Finally, the size of the market area is equal to $a^2 3\sqrt{3}/2$ and the distance b between B_1 and B_2, B_1 and $B_3, \ldots, B_1$ and B_7 is equal to $a\sqrt{3}$.

(2) Fig.39(b). The farms are located in the middle of the sides of the hexagons. The distance nV is still equal to a. Due to the competition of B_2 at A_2, B_3 at A_3, etc., B_1 supplies the equivalent of 4 farms: that is, all of A_1 plus 1/2 of the other 6 farms $A_2, A_3, \ldots, A_7$. The distance b is equal to $a\sqrt{4} = 2a$.

(3) Fig.39(c). The sides of the hexagons go through the free space between the farms. B_1 then supplies the 7 farms $A_1, A_2, \ldots, A_7$. The distance nV is still equal to a, but the distance b is equal to $a\sqrt{7}$.

Figure 39

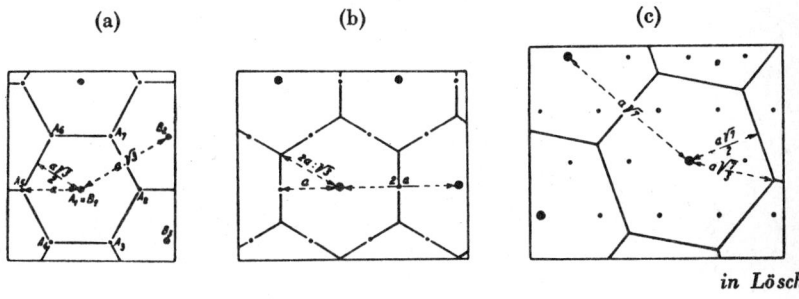

(a)   (b)   (c)

in Lösch

These three typical orientations exist simultaneously. Figure 40 shows how, in the case of 10 areas, the locations of concurrent centers are codetermined. Table 4 indicates for each area the values of n (number of centers entirely supplied, including the center of production), b and nV.

From Table 4 can be derived a simple relation between n and b, that is: $b = a\sqrt{n}$

The distance between two homogeneous firms is equal to the distance between the settlements supplied multiplied by the square root of their number. The network of market areas is extended as this operation is repeated starting from new hexagons. The upper portion of the Figure 40 partially summarizes the results then obtained. Each point represents original equidistant points. The points surrounded by a circle are supply centers. Each number indicates the dimension of the neighboring market area taking as a point of reference the size of the initial area (indicated at the center of the Figure). Those numbers which are between parentheses indicate alternative centers of supply. Note then the rise of the so-called sectors "rich in towns" (hatched zones) which are characterized by a large accumulation of supply centers.

Figure 40

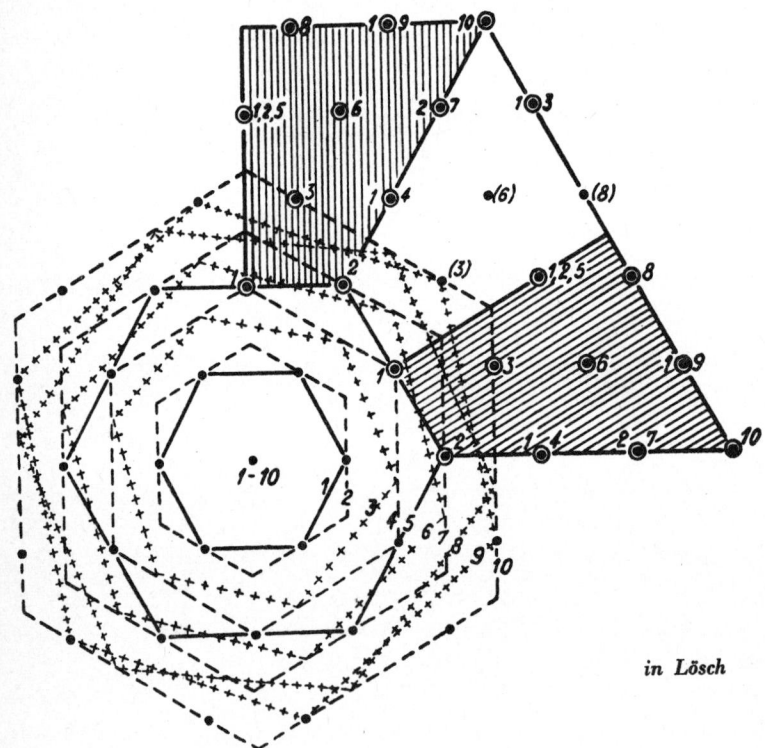

in Lösch

Table 4

Area N°	n	b	nV
1	3	$a\sqrt{3}$	a
2	4	$a\sqrt{4}$	a
3	7	$a\sqrt{7}$	a
4	9	$a\sqrt{9}$	$a\sqrt{3}$
5	12	$a\sqrt{12}$	$2a$
6	13	$a\sqrt{13}$	$a\sqrt{3}$
7	16	$a\sqrt{16}$	$2a$
8	19	$a\sqrt{19}$	$2a$
9	21	$a\sqrt{21}$	$a\sqrt{7}$
10	25	$a\sqrt{25}$	$a\sqrt{7}$

in Lösch

The complete model follows graphically. First of all, these networks of market areas, specific to each good, order themselves in a system of networks when the networks of several heterogeneous goods are combined. It suffices to arrange the networks in such a manner that they all have at least one common center (see Figure 41(a)). The networks can thus be pivoted so as to obtain six sectors rich in towns and six sectors poor in towns. (see Figure 41(b) where the networks have been eliminated).

Figure 41

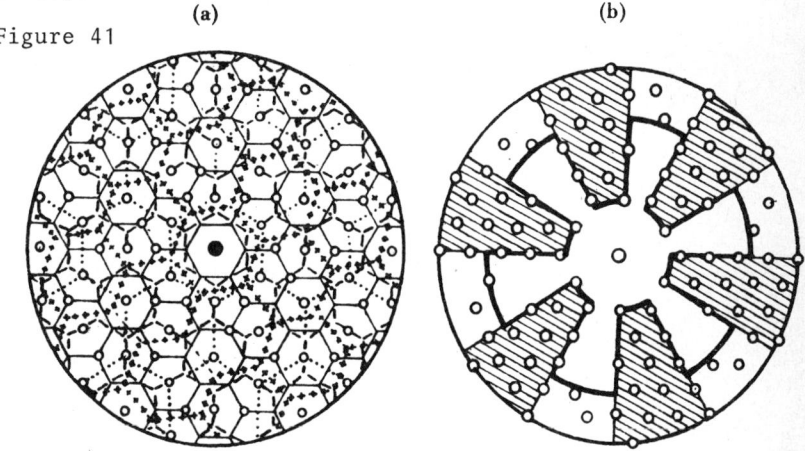

in Lösch

The system of networks which results determines the optimal lines of communications which link the regional centers and whose importance varies as a function of the number of supply centers located at each settlement (see Figure 42 on which the number of centers is indicated).

Figure 42

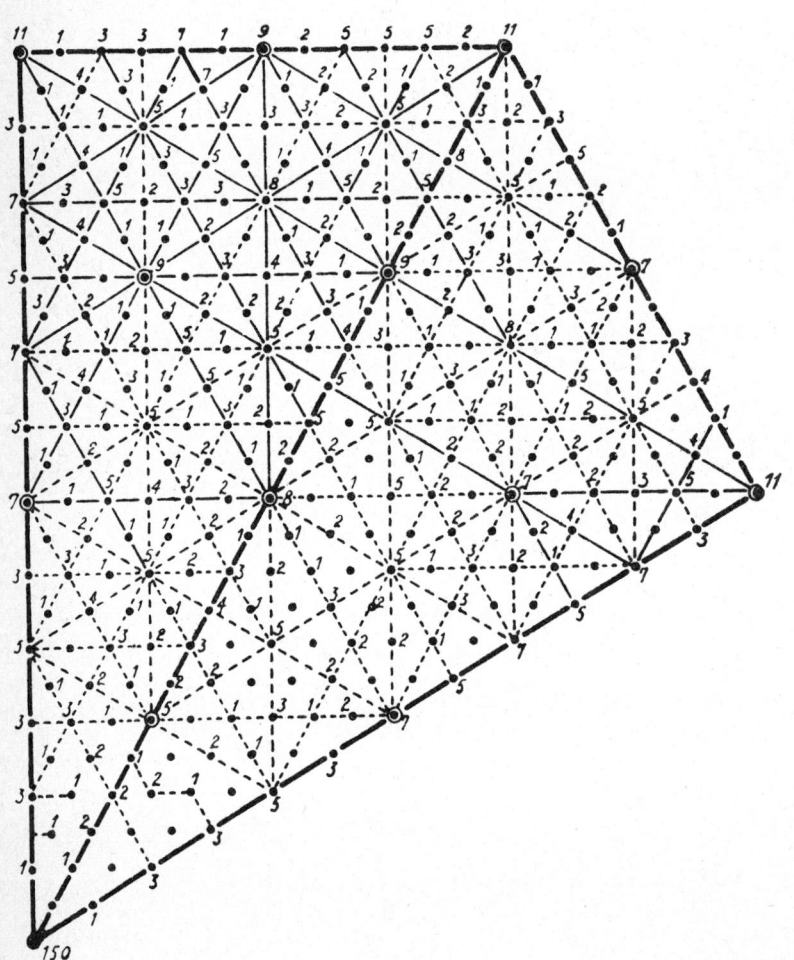

in Lösch

Let L be the radius of a system of networks. Once again, other systems appear starting from the located centers H_2, H_3, etc., at a distance equal to 2L around the center H_1 of the system considered. The number of such systems depends on the good which possesses the largest minimum sales radius (equal to or near L).

Finally, these systems tend to order themselves in networks of systems (see Figure 43(a) and (b)). When they are numerous enough, the systems of networks appear distributed in hexagonal shape. But the respective positions of the sectors rich in towns are indeterminate. In Figure 43(a) the communication lines traverse the richer territories but this is not the case for Figure 43(b). For this reason, the first assumption is more likely to be realized in practice.

Figure 43

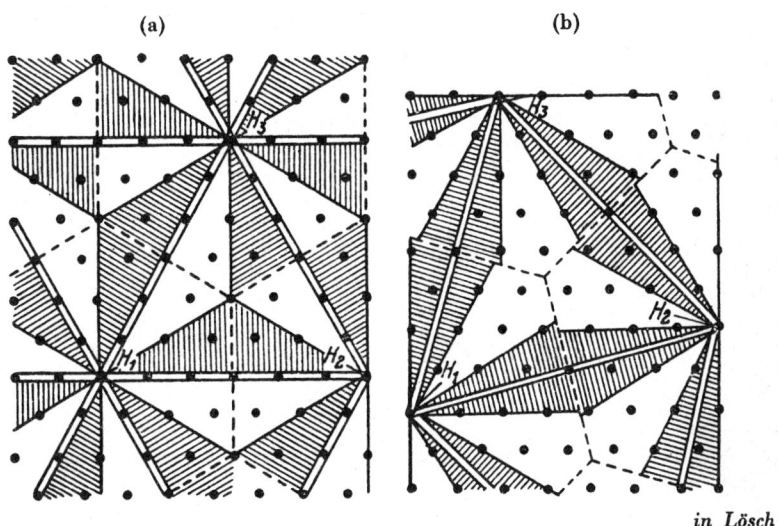

in Lösch

4 - Spatial Differentiation of Prices.

The seller has the choice between adapting his prices to the individual buyer (Case A), keeping his prices rigidly fixed so that all buyers pay the same f.o.b. price (Case F), or keeping a uniform delivered price for all the buyers (Case C).

Let d_0 be the individual demand as a linear function of the uniform delivered price (see Figure 44). The demand curve is the same for all buyers since it relates to the local price paid by each. The same is not true for the individual demand d_1, a function of the f.o.b. price, located at a distance i from d_0 (expressed in unit transport cost), and which is valid only for customers located at a constant distance from the seller.

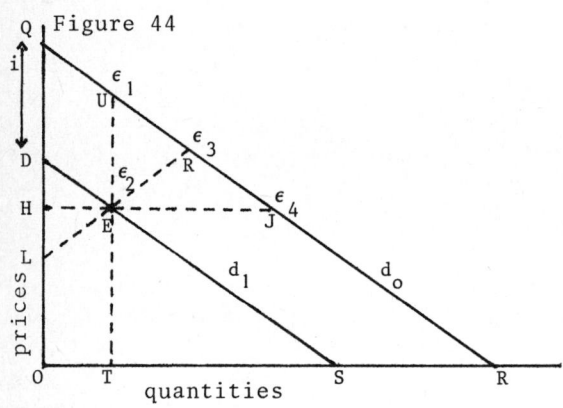

Figure 44

in Lösch

For a given price, the vertical line UT intersects d_0 at the level of the uniform delivered price and d_1 at the level of the f.o.b. price. The horizontal line, HJ, intersects d_0 and d_1 at the level of the f.o.b. price. The corresponding delivered price then lies on d_0 vertically above the intersection with d_1. A straight line, KL cuts UJ in the middle on d_0 and d_1 at the level of the f.o.b. price. Consequently, the seller pays the whole freight in Case C, one half in Case A and none in Case F. At the point U, the elasticity of the demand is $\epsilon_1 = \frac{UR}{UQ}$; at the point E, $\epsilon_2 = \frac{ES}{ED}$; at the point K, $\epsilon_3 = \frac{KR}{KQ}$; and at the point J, $\epsilon_4 = \frac{JR}{JQ}$.

But since UQ = ED and UR > ES, $\epsilon_1 > \epsilon_2$. Corresponding comparisons give: $\epsilon_2 > \epsilon_3 > \epsilon_4$. The individual demand with respect to the f.o.b. price is less elastic than the individual demand with respect to the corresponding uniform delivered price. To proceed from individual demand to total demand requires the assumption that the demand on n localities is equal to n times the demand of one. But, except for Case C, the dispersion of n demands makes the total demand of the area smaller and more elastic than the local demand.

The search of the optimal price by the seller leads him to fix an f.o.b. price such that the marginal revenue equals the marginal cost. The most satisfactory situation is the one for which this condition is satisfied for each buyer: that is, in Case A.

Now let:

t = the freight cost per unit from factory to buyer
P = the delivered price [P = f(n)]
p = the factory price [p = P-t]
n = the individual demand at delivered price
$\epsilon = -\frac{P \cdot dn}{n \cdot dP}$ the elasticity of demand in respect to delivered price
c = the marginal costs

Marginal revenue is equal to:

$$\frac{d(Pn)}{dn} - t = \frac{Pdn + ndP}{dn} - t = P - \frac{P}{\epsilon} - t = P\left(\frac{\epsilon - 1}{\epsilon}\right) - t$$

Marginal revenue equals marginal costs when:

$P\left(\frac{\epsilon-1}{\epsilon}\right) - t = c$: that is, when $P = \frac{\epsilon(c+t)}{\epsilon-1}$

Thus: $p = P - t = \frac{\epsilon(c+t)}{\epsilon-1} - t = \frac{\epsilon c + t}{\epsilon-1}$

For each buyer, p will depend thus on t, ϵ, and c.
Then:

(1) p rises with t if ϵ and c are constant. Generally ϵ rises also with t. But p rises with t, and diminishes when ϵ rises. Thus the question of whether p rises with distance depends on the relationship between t and ϵ.

Certain conditions must be fulfilled in order for p to rise with distance. Let A_1 be nearer to the factory than A_2, p_2 will then be greater than p_1 if:

$$\frac{\epsilon_2 c + t_2}{\epsilon_2 - 1} > \frac{\epsilon_1 c + t_1}{\epsilon_1 - 1}$$

From this it follows that: $\frac{\epsilon_2(t_1+c) + (t_2-t_1)}{\epsilon_1(t_2+c)} < 1$

But $\frac{t_2-t_1}{\epsilon_1(t_2+c)}$ is necessarily smaller than 1, and the other terms in the inequality are smaller than 1 if $(\epsilon_2 t_1 + \epsilon_2 c) < (\epsilon_1 t_1 + \epsilon_1 c)$. Nevertheless $\epsilon_2 c$ is generally larger than $\epsilon_1 c$. The inequality supposes that t rises more proportionally than ϵ, which appears improbable since ϵ can become infinite while t always remains finite. If linear demand curve as the average shape is assumed, t increases less rapidly than ϵ. Consequently, p will have a tendency to fall when distance rises.

(2) The more elastic the demand, the lower is the factory price p that must be paid. At the limit, p tends to coincide with c; it can even be lower than c if $\epsilon < 1$. The equilibrium point thus falls in the elastic portion of the demand curve.

(3) The higher the marginal cost c, the higher is the factory price p which the same buyer must pay. An increase in c affects p more for the farthest buyers than for the buyers who are closest to the place of production, thus resulting in a reduction of spatial price differences. The converse is true for a fall in c.

Chronological Bibliography

This bibliography is limited to works of abstract analysis and directly related topics.

The most frequent references are abbreviated as follows:

Amer.Econ.R.	The American Economic Review
Econ.J.	The Economic Journal
Jb.Nat.Oekon.Statist.	Jahrbücher für Nationalokonomie und Statistik
J.Polit.Econ.	The Journal of Political Economy
Jb.Sozialwissens.	Jahrbuch fur Sozialwissenschaft
J.R.S.	Journal of Regional Science
Papers R.S.A.	Papers, Regional Science Association
Quart.J.Econ.	The Quarterly Journal of Economics
R.Econ.Statist.	The Review of Economics and Statistics
R.Econ.Stud.	The Review of Economic Studies
R.S.U.E.	Regional Science and Urban Economics
R.U.E.	Regional and Urban Economics. Operational Methods.
Schmollers Jb.Gesetzgebg.	Schmollers Jahrbuch fur Gesetzgebung,
Verw.Volksw.	Verwaltung und Volkswirtschaft
Weltwirts.Archiv.	Weltwirtschaftliches Archiv.

The references are given in chronological order. For any given year, they are classed in alphabetical order.

BEFORE 1900

[1] CANTILLON (R.): <u>Essai sur la nature du commerce en general.</u> Written 1725, ed.1755, reed.Paris, I.N.E.D., 1952. English trans. by HIGGS (H.), London, Macmillan, 1931.

[2] STEUART (J.): <u>An Inquiry into the Principles of Political   Economy.</u> London, 1767, French trans.Paris, 1789. English reed.by SKINNER (A.S.), Chicago, University Press, 1966.

[3] CONDILLAC (Abbe De): <u>Le Commerce et le Gouvernement consideres relativement l'un a l'autre.</u> Amsterdam, Paris, Jombert et Cellot, 1776.

[4] SMITH (A.): <u>An Inquiry into the Nature and Causes of the Wealth of Nations.</u> London, 1776, French trans.Paris, Garnier, 1843.

[5] THUNEN (J.H.Von): <u>Der isolierte Staat in Beziehung auf Landwirtschaft und Nationalokonomie.</u>
Part I, 1st ed., Hamburg, Perthes, 1826, 2nd ed., Rostock, Leopold , 1842, French trans.by LAVERRIERE (J.), Paris, Guillaumin, 1851.
Part II, Sect.1, Rostock, Leopold, 1850, French.trans.by WOLKOFF, Paris, Guillaumin, 1857.
Part II, Sect.2, and Part III, Posthumous ed., Rostock Leopold, 1863.
Part I and Part II, Sect.1, reed.by WAENTIG (H.), Jena, G.Fischer, Coll.Sammlung sozialwissenschaftlicher Meister, 13, 1st ed. 1910, 2nd ed. 1921, 3rd ed. 1930.
<u>Thunen, Der isolierte Staat,</u> Darmstadt,Wissenschaftliche Buchgesellschaft, 1966.
Partial English trans. by WARTENBERG (C.W.): <u>Von THUNEN'S Isolated State,</u> HALL (P.)ed., Oxford, Pergamon Press,1966.

[6] URE : <u>The Philosophy of Manufactures,</u> 1835.

[7] ROSCHER (W.): <u>Ideen zur Politik und Statistik der Ackerbausysteme. Archiv der politischen Okonomie und Polizeiwissenschaft,</u> 8, 1845, p.173ff.

[8] KOHL (J.G.): <u>Der Verkehr und die Ansiedelungen der Menschen in ihrer Abhangigkeit von der Gestaltung der Erdoberflache.</u> Leipzig, 2nd ed., 1850.

[9] ROSCHER (W.): <u>Studien uber die Naturgesetze, Welche den Zweckmassigen Standort der Industriezweige bestimmmen.</u> Ansichtsen der Volkswirtschaft aus dem geschichtlichen Standpunkte. 1st ed. 1865, 3rd ed. 1878.

[10] SCHUMACHER-ZARCHLIN (H.): <u>Johann Heinrich von Thunen. Ein Forscherleben.</u> Rostock, 1868.

[11] LAUNHARDT (W.): <u>Die Kommerzielle Trassierung der Verkehrswege.</u> Hannover, Gehmorl und von Geefeld, 1872.

[12] LAUNHARDT (W.): Kommerzielle Trassierung der Verkehrswege. <u>Zeitschrift fur Architektur und Ingenieurswesen,</u> 18,1872, 515-534.

[13] SCHAFFLE (A.): <u>Das Gesellschaftlichen System der menschlichen Wirtschaft.</u> Tubingen, 3rd ed., 1873.

[14] DAVIDSON (O.): <u>Bitrag till jordrantenteories historia.</u> Uppsala, Universitetets Arsskcrift, 1880.

[15] LAUNHARDT (W.): Die Bestimmung des zweckmassigsten Standortes einer gewerblichen Anlage. <u>Zeitschrift des Vereins Deutscher Ingenieure,</u> 26, 1882, 106-115.

[16] LAUNHARDT (W.): <u>Mathematische Begrundung der Volkswirtschaftslehre.</u> Leipzig, B.G.Teubner, 1885.

17] LAUNHARDT (W.): Theorie des Trassierens, Hanover,2 vol.,1887-1888. English trans. by BEWLEY (A.): The Principles of Railway Location, Madras, Lawrence Asylum Press, 2 vol.,1900-1902.
[18] MOORE (H.L.): Von Thunen's Theory of Natural Wages. Quart.J.Econ., 1895, p.291ff.
[19] ROSS (E.A.):The Location of Industries. Quart.J.Econ., 1896,p.246ff.

FROM 1900 TO 1920

[20] HALL (F.S.): The Localization of Industry. The Twelfth Census of the United States, 39, 1900, Vol.7, Manufactures I.
[21] CLARK (J.B.): Control of Trusts, 1901, ed.rev.by J.B.& J.M.CLARK,1912
[22] CUNNINGHAM : The Localization of Industry. Econ.J. 1902.
[23] AEREBOE (F.): Beitrage zur Wirtschaftslehre der Landbaus. 1905.
[24] PASSOW (Th.): Die Methode der nationalokonomischen Forschungen J.H. von Thunen. Tubingen, 1905.
[25] MAUNIER (R.): La distribution geographique des industries. Revue Internationale de Sociologie. 1908.
[26] WATERSTRADT : Die Rentabilitat der Wirtschafts-systeme nach J.H. von Thunen isoliertem Staat und in unserer Zeit. 1909.
[27] WEBER (A.) : Ueber den Standort der Industrien, Part.I: Reine Theorie des Standorts. Tubingen, 1909 (1st ed.), 1922 (2d ed.). English translation by FRIEDRICH (C.J.): Alfred Weber's Theory of Location of Industries. Chicago, University of Chicago Press, 1st ed., 1929,2d ed.1957.
[28] BORTKIEWICZ : Eine geometrische Fundierung der Lehre vom Standort der Industrien. Archiv fur Sozialwissenschaft und Sozialpolitik, 30,1910,769-771.
[29] MAUNIER (R.): Theories sur la formation des villes. Revue d' Economique Politique, 1910,546-560 and 637-655.
[30] MAUNIER (R.): L'origine etla fonction economique des villes. Etude de morphologie sociale. Paris, Bibliotheque Sociologique Internationale, 42, 1910.
[31] SOMBART (W.): Einige Bemerkungen zur Lehre von Standort der Industrien. Archiv fur Sozialwissenschaft und Sozialpolitik, May 1910, 748-758.
[32] WEBER (A.): Die Standortslehre und die Handelspolitik. Archiv fur Sozialwissenschaft und Sozialpolitik. 32, 1911,667-688.
[33] AUERBACH (F.): Das Gesetz der Bevolkerungskonzentration. Petermanns Mitteilungen, 59-1,1913,74-77.
[34] FURLAN (V.): Die Standortsprobleme in der Volks und Weltwirtschaftslehre. Weltwirts.Archiv. 2,1913, 1-34.
[35] WICKSELL : Vorlesungen uber Nationalokonomie. Jena,1913,1,p.176ff.
[36] WEBER (A).: Industrielle Standortslehre: Reine und Kapitalistische Theorie des Standorts. Grundriss der Sozialokonomik, 6, 1914, 54-82.
[37] FETTER (F.A.): Economic Principles. New-York, 1915.

FROM 1921 TO 1930

[38] KEIR (M.): Economic Factors in the Location of Manufacturing Industries. Annals of the American Academy of Political and Social Sciences. 1921.
[39] BRINKMANN (T.): Die Oekonomik des landwirtschaftlichen Betriebes. Tubingue, Grundriss der Sozialokonomik, 1922. English trans.by BENEDICT (E.T.), STIPPLER (H.) and BENEDICT (M.R.): Theodor Brinkmann's Economics of the Farm Business. Berkeley, University of California Press,1935.
[40] AEREBOE (F.): Allgemeine landwirtschaftsliche Betriebslehre. Berlin P.Parey, 1923.
[41] ENGLANDER (O): Volkswirtschaftliche Theorie des Personenverkehrs. Archiv fur Sozialwissenschaften und Sozialpolitik, 50,1923,p.653ff.
[42] DOBBELER (C.von): Mathematische Beitrage zur Wirtschafsgeographie. Technik und Wirtschaft, 1924.
[43] ENGLANDER (O.): Theorie des Guterverkehrs und der Frachtsatze. Jena, G.Fischer, 1924.
[44] FETTER (F.A.): The Economic Law of Market Areas. Quart.J.Econ. 38,1924, p.525ff.
[45] OHLIN (B.): Handelns Teori. Stockholm, 1924.
[46] ROSCHER (W.): Geschichte der National-Okonomik in Deutschland. Geschichte der Wissenschaften in Deutschland, Munchen, Berlin, 1924.
[47] SCHILLING (A.): Die Wirtschaftsgeographischen Grundgesetze des Wettbewerbs in mathematischer Form. Technik und Wirtschaft. 17,1924,p.145ff.
[48] SCHNEIDER (E.):Mathematische Betrachtungen uber den nationalen Gutertransport. Technik und Wirtschaft, 17,1924,p.204ff.
[49] HAWTREY (R.G.): The Economic Problem. New-York, 1925, London, 1926, chapters 7 and 9.
[50] LOTKA (A.J.): Elements of Physical Biology. Baltimore, Williams and Wilkins, 1925.
[51] PREDOHL (A.):Das Standortsproblem in der Wirtschaftstheorie. Weltwirts.Archiv, 21,1925, 294-331.
[52] SCHMIDT (P.H.): Wirtschaftsforschung und Geographie. Jena, 1925.
[53] BLACK (J.D.): Introduction to Production Economics. New-York, Henry Holt and Co., 1926.
[54] ENGLANDER (O.): "Standort" in Handworterbuch der Staatswissenschaften. Jena,1926.
[55] ENGLANDER (O.): Kritisches und Positives zu einer allgemeinen reinen Lehre vom Standort. Zeitschrift fur Volkswirtschaft und Sozialpolitik. Neue Folge, 5, 1926, 474-479.
[56] SRAFFA (P.): The Laws of Returns Under Competitive Conditions. Econ.J. Dec.1926, 535-550.
[57] WEIGMANN (H.): Kritischer Beitrag zur Theorie des internationalen Handels. G.Fischer. Jena, 1926.
[58] KRZYZANOWSKI (W.): Review of the Literature on the Location of Industries. J.Polit.Econ. 35, 1927, 278-291.
[59] NAPP-ZINN (A.F.): Zur Theorie des Standorts. Zeitschrift fur Verkehrswissenschaft, 5,1927.
[60] PREDOHL (A.): Zur Frage einer allgemein Standortstheorie. Zeitschrift fur Volkswirtschaft und Sozialpolitik, 5, 1927, 756-763.
[61] RITSCHL (H.):Reine und historische Dynamik des Standortes der Erzengungszweige. Schmollers Jb.Gesetzgebg. Verw.Volksw. 51, 1927, 813-870.

[62] KRZYMOWSKI (R.): Graphical Presentation of Thunen's Theory of
            Intensity. Journal of Farm Economics, 10,1928,461-482.
[63] PREDOHL (A.): The Theory of Location in its Relation to General
            Economics. J.Polit.Econ., 36,1928, 371-390.
[64] ENGLANDER (O.): Preisbildung und Preisaufbau. Theorie des
            Volkswirtschaft, 1929.
[65] HOTELLING (H.): Stability in Competition. Econ.J., 39, 1929,41-57.
[66] DAHMEN (J.): Die Aachener Tuchindustrie. Ihre Wirtschaftlichen und
            sozial grundlagen. Eine Untersuchung der
            Standortsfaktorem. Leipzig,1930.
[67] JONASSON (O.): Jordburkets beroende av det geografika marknadslaget
            Kungliga Lantbruks akademiens Handlingar. 1930.
[68] LOSCH (A.):Eine Auseinandersetzung uber das Transferproblem.
            Schmollers Jb.Gesetzgegb.Verw.Volksw., 54-55,1930.

FROM 1931 TO 1935

[69] DIEMER (H.): Industrial Organisation and Management, Chicago,
            1931. Chap.III. Locating an Industry.
[70] FETTER (F.A.): The Masquerade of Monopoly, New-York, Harcourt,
            Brace & Co Inc., 1931.
[71] REILLY (W.J.): The Law of Retail Gravitation. New-York,
            Knickerbrocker Press, 1931.
[72] ROBINSON (E.A.G.): The Structure of Competitive Industry, Cambridge
            Economic Handbooks, London, 1931-45.
[73] WEIGMANN (H.): Ideen zu einer Theorie der Raumwirtschaft.
            Weltwirts.Archiv., 34,1931, 1-40.
[74] PFANNSCHMIDT (M.): Standort, Landesplanung, Baupolitik. Berlin
            1932
[75] CHAMBERLIN (E.H.): The Theory of Monopolistic Competition.
            Cambridge Harvard University Press, 1933, 6e ed. 1948 and
            Doctoral Dissertation, id.,Harvard University
            Library.1927.
[76] CHRISTALLER (W.): Die zentralen Orte in Suddeutschland: Eine
            Okonomisch-geographische Untersuchung uber die
            Gesetzmassigkeit der Verbreitung und Entwicklung der
            Siedlungen mit stadtischen Funktionen. Jena, Gustav
            Fischer Verlag, 1933. English trans. by BASKIN (C.W.):
            Central Places in Southern Germany, New Jersey,
            Prentice-Hall,Inc., Englewood Cliffs, 1966.
[77] OHLIN (B.): Interregional and International Trade. Cambridge
            Harvard University Press, 1933.
[78] WEIGMANN (H.): Standortstheorie und Raumwirtschaft. Betrachtungen
            zur Entwicklung der Standortstheorie in Deutschland seit
            John-Heinr.von Thunen. W.Seedorf-H.Jurgen: John-Heinr.
            von Thunen zum 150 Geburtstag. Rostock, Carl Hinstorffs,
            1933.
[79] ZEUTHEN (F.): Theoretical Remarks on Price Policy: Hotelling's Case
            With Variations. Quart J.Econ. Feb.1933, 231-253.
[80] ZEUTHEN (F.): Afstanden Mellem Bedriftscellerne ogdet Prispolitiske
            Sammenspil. Nationalokonomisk Tidsskrift, 1933, p.209ff
[81] CHRISTALLER (W.): Allgemeine geographische Voraussetzungen der
            deutschen Verwaltungsgliederung. Jahrbuch fur Kommunal-
            wissenschaft, 1,1934.
[82] HOGBOM (I.): Varldssjofarten. Dess geografiska och ekonomiska
            betingelser. Goteborg, 1934.

[83] LOSCH (A.): Selbskosten und Standortverschiebungen von Genussgutern nach dem Krieg als Ursachen von Zolltendenzen. Zwischenstaatliche Wirtschaft ed. H.V.Beckerath. Berlin, 1934.
[84] ROBINSON (A.): The Problem of Management and the Size of Firms. Econ.J. June 1934, 242-257.
[85] SCHNEIDER (E.): Preisbildung und Preispolitik unter Berucksichtigung der geographischen Verteilung von Erzengern und Verbrauchern. Schmollers Jb.Gesetzgebg.Verw. Volksw. 1934,257-277.
[86] OHLIN (B.): Some Aspects of the Theory of Rent:von Thunen vs.Ricardo. in Economics, Sociology and the Modern World: Essays in Honour of T.N.Carver. Cambridge, Mass,1935, 171-183.
[87] PALANDER (T.): Beitrage zur Standortstheorie. Uppsala, Almqvist & Wiksells Boktryckeri. A.B.,1935.
[88] SCHNEIDER (E.): Bemerkungen zu einer Theorie der Raumwirtschaft. Econometrica. Jan.1935, 79-105.
[89] TINTNER (G.): Die Nachfrage im Monopolgebiet. Zeitschrift fur Nationalokonomie, 1935, 536-539.
[90] WEIGMANN (H.): Politische Raumsordnung. Gedanken zur Neugestaltung des deutschen Lebensraumes. Hanseatische Verlagsanstalt. Hamburg, 1935.

FROM 1936 TO 1940

[91] BURNS (A.R.): The Decline of Competition .New-York, 1936.
[92] LOSCH (A.) : Bevolkerungswellen und Wechsellagen. Beitrage zur Erforschung der Wirtschaftlichen Wechsellagen Aufschwung Krise, Stockung. A.Spiethoff, 13.Jena, 1936.
[93] SINGER (H.W.) : The "Courbe des Populations": A Parallel to Pareto's Law. Econ.J. 46,1936, 254-263.
[94] CHRISTALLER (W.) : Die landliche Seidlungsweise im Deutschen Reich und ihre Beziehungen zur Gemeindeorganization. Stuttgart und Berlin, W.Kohlhammer, 1937.
[95] HOOVER (E.M.): Location Theory and the Shoe and Leather Industry. Cambridge, Mass.Harvard University Press, 1937.
[96] HOOVER (E.M.): Spatial Price Discrimination. R.Econ.Stud. ,4,1937, 182-191.
[97] LERNER (A.P.)and SINGER (H.W.): Some Notes on Duopoly and Spatial Competition. J.Polit.Econ. April 1937, 145-186.
[98] SINGER (H.W.): A Note on Spatial Price Discrimination. R.Econ.Stud. 5,1937, 75-77.
[99] SEIDLER : Geographical Price Relations and Competition. Journal of Marketing. 1937, p.198ff
[100] BARFOD (B.): Local Economic Effects of a Large-Scale Industrial Undertaking. London, 1938, Economic Research Department, Oxford University Press, New-York, 1938.
[101] CHRISTALLER (W.): Rapports fonctionnels entre les agglomerations urbaines et les campagnes. Comptes-Rendus du Congres International de Geographie. Amsterdam, 1938, 123-137.
[102] DEAN (W.H.Ir.): The Theory of the Geographic Location of Economic Activities. Selections from the doctoral dissertation. Cambridge, Mass.Harvard University Press, 1938.
[103] LOSCH (A.): Beitrage zur Standortstheorie. Schmollers Jb. Gesetzgebg.Verw.Volksw. 62,1938,329-335.
[104] LOSCH (A.): The Nature of Economic Regions. The Southern Economic Journal, 5,1938,71-78.

[105] LOSCH (A.): Wo gilt das Theorem der Komparitiven Kosten? Weltwirts. Archiv. 48,1938,45-64.
[106] STACKELBERG (H.Von): Das Brechungsgesetz des Verkehrs. Jahrb.f. Nat. und Stat. 148,1938,680-696.
[107] CLARK (J.M.): One Form of Price Competition between Geographically Separated Large Producers. Econometrica. 1939,174-177.
[108] DENNISON (S.R.): The Location of Industry and the Depressed Areas. Oxford University Press, 1939.
[109] LOSCH (A.): Eine neue Theorie des internationalen Handels. Weltwirts.Archiv., 50,1939,308-327.
[110] SINGER (H.W.): Regional Labour Markets and the Process of Unemployment. R.Econ.Stud. 7,1939,42-58.
[111] COPELAND (M.): Competing Products and Monopolistic Competition. Quart.J.Econ. 55,1940-41,1-35 and 3-23.
[112] LOSCH (A.): Die raumliche Ordnung der Wirtschaft. Jena.Gustav Fischer.1st ed.1940;2nd ed.1944;3rd ed.1962; English trans.by WOGLOM (W.H.) and STOLPER (W.F.): The Economics of Location. New Haven, Yale University Press, 1954. Spanish trans. by ARNOLD (G.H.) and CASSENS (F.): Teoria Economica Espacial. Buenos-Aires,Editorial "El Ateneo" 1957.
[113] PRIBRAM (K.): Residual, Differential and Absolute Urban Ground Rents and Their Cyclical Fluctuations. Econometrica. Jan.1940,62-78.
[114] ROPKE (W.): L'emplacement de la production, in L'Explication economique du monde moderne. French.trans.by BASTIER (P.) Paris, Medicis, 1940-45, 122-24.
[115] STOUFFER (S.A.): Intervening Opportunities: A Theory Relating Mobility and Distance. American Sociological Review, 5,1940,845-867; Republished in STOUFFER (S.A.): Social Research to Test Ideas, chap.4,68-91,New-York, The Free Press of Glencoe, 1962.
[116] TOSCHI (M.U.): Considerations sur la geographie economique en tant que science economique. Etudes en l'honneur de Louis Amoroso. Annales de l'Institut de Statistiques de l'Universite de Bari, 18,1940.
[117] ULLMAN (E.L.): A Theory of Location for Cities. The American Journal of Sociology, 46,1940-1941,853-864.

FROM 1941 TO 1945

[118] CHRISTALLER (W.): Raumtheorie und Raumordnung. Archiv fur Wirtschaftsplanung, 1,1941,122-126 and 131-133.
[119] HITCHCOCK (F.): The Distribution of a Product from Several Sources to Numerous Localities. Journal of Mathematics and Physics, 20,1941,224-230.
[120] MEYER (F.): Eine neue Transfertheorie? Archiv fur Wirtschaftspl. 1941.
[121] LOSCH (A.): Die Lehre vom Transfer neugefaszt. Jb.Nat.Oekon. Statist. 154,1941.
[122] LOSCH (A.):Die Leistung der Seeschiffahrt. Nauticus, 1941,326-336.
[123] RITSCHL (H.): Aufgabe und Methode der Standortslehre. Weltwirts. Archiv. 53,1941,115-125.
[124] ROBINSON (A.): A Problem in the Theory of Industrial Location. Econ.J., 1941,270-75.
[125] SCHNEIDER (E.): Der Raum in der Wirtschaftstheorie. Jahrb.Nat. Oekon. Statist. 1941.
[126] SMITHIES (A.): Monopolistic Price Policy in a Spatial Market. Econometrica, Jan.1941,63-73.

[127] SMITHIES (A.): Optimum Location in Spatial Competition. J.Polit. Econ., June 1941, 423-439.
[128] VOELCKER (A.): Matz und Zahl in der Raumforschung. Gustav Fischer. Jena, 1941.
[129] ZIPF (G.K.): National Unity and Disunity. The Nation as a Bio-Social Organism. Bloomington, Indiana, 1941.
[130] ACKLEY (G.): Spatial Competition in a Discontinuous Market. Quart. J.Econ., Feb.1942,212-230.
[131] ENKE (S.): Space and Value. Quart.J.Econ. August 1942,627-637.
[132] SMITHIES (A.): Aspects of the Basing-Point System. Amer.Econ.R. 32,1942,423-440.
[133] ACKLEY (G.): Price Policies in Industrial Location and Natural Resources, National Resources Planning Board.1943.
[134] ALLAIS (M.): A la recherche d'une discipline economique. Paris, Imprimerie Nationale, 1943.
[135] CLARK (J.M.): Imperfect Competition Theory and the Basing-Point Problem. Amer.Econ.R., June 1943.
[136] MOLLER (H.): Die Formen der regionalen Preisdifferenzierung. Weltwirts.Archiv., 1943, p.81ff.
[137] MOLLER (H.): Grundlagen einer Theorie der regionalen Preisdifferenzierung. Weltwirts.Archiv. 58,1943,335-390.
[138] LOSCH (A.): Um eine neue Transfertheorie: Zur Verteidigung der alten Lehre durch Fritz Meyer. Jb.Nat.Oekon.Statist., 1943.
[139] NIEDERHAUSER (E.): Die Standortstheorie Alfred Webers. Staatswissenschaftliche Studien, 14, Weinfelden, 1944.
[140] CLARK (C.): The Economic Functions of a City in Relation to Its Size. Econometrica. 13,1945,p.97ff.
[141] LEWIS (W.A.): Competition in Retail Trade. Economica, 12, 1945, p.202ff.
[142] McLAUGHLIN (G.E.): Industrial Expansion and Location. Annals of the American Academy of Political and Social Science. Nov.1945.
[143] VINING (R.): Regional Variation in Cyclical Fluctuation Viewed as a Frequency Distribution. Econometrica. July 1945, 183-213.

FROM 1946 TO 1950

[144] REYNOLDS (L.G.): Wage Differences in Local Labor Markets. Amer. Econ.R., 36,1946,366-375.
[145] TOSCHI (M.U.): Considerations et recherches sur la localisation des industries en fonction des moyens de transport. Annales de la Faculte d'Economie et de Commerce, Bari, 1946.
[146] TOSCHI (M.U.): Localisation des industries et inertie. L'actualite econ.et fin.a l'etranger. April 1947,5-12, trans. Giornale degli Economisti,Sept-Oct.1946.
[147] TOSCHI (M.U.): La theorie de la localisation des industries selon A.Weber. Institut de Geographie de l'Universite de Bari. Memoires, 9, Ed.Macri.Bari.1946.
[148] VINING (R.): Location of Industry and Regional Patterns of Business-Cycle Behavior. Econometrica. Jan.1946,37-68.
[149] VINING (R.): The Region as a Concept in Business-Cycle Analysis. Econometrica. July 1946,201-218.

[150] EITEMAN: Factors Determining the Location of the Least Cost Point. Amer.Econ.R. 1947, p.910ff.
[151] VINING (R.): Measuring State and Regional Business Cycles. J.Polit.Econ. 55,1947,p.346ff.
[152] EDWARDS (C.D.): The Effect of Recent Basing Point Decisions Upon Business Practices. Amer.Econ.R., 38,1948,828-842.
[153] FETTER (F.A.): Exit Basing Point Pricing. Amer.Econ.R., 38,1948, 815-827.
[154] HOOVER (E.M.): The Location of Economic Activity. New-York, Mc.Graw-Hill Book Company, Inc.1948. Partial French trans.by ALAURENT (J.):La localisation des activites economiques, Les Editions Ouvrieres, Paris,1955.
[155] NOYES (C.R.): Economic Man in Relation to His Natural Environment, The Columbia University Press. New-York,1948,vol.11, 695-1443.
[156] SAUVY (A.): Migrations et commerce international. Faut-il transporter les hommes ou les marchandises? Economia Internazionale, August 1948,803-812.
[157] CLARK (J.M.): The Law and Economics of Basing Points: Appraisal and Proposals. Amer.Econ.R. ,39,1949,430-447.
[158] CONVERSE (P.D.): New Laws of Retail Gravitation. The Journal of Marketing, 14,1949,379-384.
[159] GIERSCH (H.): Economic Union Between Nations and the Location of Industries. R.Econ.Stud., 1949-1950, 17,87-97.
[160] ISARD (W.): The General Theory of Location and Space-Economy. Quart.J.Econ. ,63,1949,476-506.
[161] KOOPMANS (T.C.): Optimum Utilization of the Transportation System. Econometrica. 17,1949, Suppl.136-146.
[162] LOSCH (A.):Theorie der Wahrung. Weltwirts.Archiv., 62,1949,p.35ff.
[163] NEFF (P.) Interregional Cyclical Differentials: Causes,Measurement and Significance. Amer.Econ.R. 39,1949,105-119.
[164] PREDOHL (A.)): Aussenwirtschaft, Weltwirtschaft, Handelspolitik und Wahrungspolitik. Grundriss der Sozialwissenschaft. Gottingen-Vandenhoeck & Ruprecht, 1949.
[165] STRODTBECK (F.):Equal Opportunity Intervals: A Contribution to the Method of Intervening Opportunity Analysis. American Sociological Review, 14,1949,490-497.
[166] VINING (R.): The Region as an Economic Entity and Certain Variations to Be Observed in the Study of Systems of Regions. Amer.Econ.R., 39,1949,89-104.
[167] ZIPF (G.K.): Human Behavior and the Principle of Least Effort. Cambridge, Addison-Wesley Press, 1949.
[168] BECKMANN (M.J.): A Formal Approach to Localization Theory. Cowles Comm.Discussion Papers: Economics n.293, 1950.
[169] BOGUE (D.J.): The Structure of the Metropolitan Community. A Study of Dominance and Subdominance. University of Michigan Institute for Human Adjustment, 1950.
[170] BRAEUER (W.Von): Der Mathematiker - Okonom.Zur Erinnerung an J.H.von Thunen. Kyklos, 4,1950,150-171.
[171] BRAEUER (W.Von): Thunen et la France. Revue d'Histoire Economique et Sociale, 28,1950,186-191.
[172] BULOW (F.): Thunen als Raumdenker. Welwirts.Archiv., 65,1950,1-24.
[173] CHRISTALLER (W.):Das Grundgerust der raumlichen Ordnung in Europa: Die Systeme der europaschen zentralen Orte. Frankfurter Geographische Hefte, 24,1950,5-14.
[174] HOFFMANN (F.): J.H.Von Thunen im Blickfeld des deutschen Kameralismus. Weltwirts.Archiv., 1,1950,25-40.
[175] HYSON (C.D.) and HYSON (W.P.): The Economic Law of Market Areas. Quart.J.Econ. 64,1950,319-327.

[176] LEONTIEF (W.W.) and ISARD (W.):Regional Applications of Input-Output Analysis. American Winter Meeting of the Econometric Society. Chicago, december 27-30, 1950.
[177] PREDOHL (A.): Weltwirtschaft in raumlicher Perspektive. Economia Internazionale. 3,1950,1044-1065.

FROM 1951 TO 1955

[178] BECKMANN (M.J.): Eine Note "Zur Theorie des raumlichen Gleichgewichts". Weltwirts.Archiv. 67,1951,167-168.
[179] BOGUE (D.J.):State Economic Areas: A Description of the Procedure Used in Making a Functional Grouping of Counties of the United States. Washington, 1951.
[180] BRAEUER (W.Von): Johann Heinrich von Thunen. Ausgewahlte Texte. Meisenheim Glan, Westkulturverlag Anton Hain,1951.
[181] CHAMBERLIN(E.H.):Monopolistic Competition Revisited. Economica, 18,1951,343-362.
[182] DANTZIG (G.B.): Application of the Simplex Method to a Transportation Problem. Chap.23 in Activity Analysis of Production and Allocation, 359-373, Cowles Comm.Monograph.13. T.C.Koopmans ed.,New-York,John Wiley and Sons, 1951.
[183] DZIEWONSKI (K.):Zagadnienia lokalizacji produkcji. Warszawa, 1951.
[184] ENKE (E.): Equilibrium Among Spatially Separated Markets:Solution by Electric Analogue. Econometrica. Jan.1951,40-47.
[185] FISCHER (E.): Urban Real Estate Markets: Characteristics and Financing. N.B.E.R.,New-York,1951.
[186] FREUTEL (G.): Industrial and Regional Structure of the American Economy. Introduction. Washington University International Economics Research Project. June 1951.
[187] GREENHUT (M.L.): Observations of Motives to Industry Location. The Southern Economic Journal, 18, 1951,225-228.
[188] ISARD (W.): Distance Inputs and the Space Economy. Part.I. The Conceptual Framework. Part.II. The Locational Equilibrium of the Firm. Quart.J.Econ. ,65,1951,181-198 and 373-399.
[189] ISARD (W.) and FREUTEL (G.): Regional and National Product Projections and Their Interrelations. in "Long-Range Economic Projection". Studies in Income and Wealth, 16, Conference on Research in Income and Wealth, New-York, N.B.E.R.,1951, Princeton University Press,1954,427-471.
[190] ISARD (W.): Interregional and Regional Input-Output Analysis: A Model of a Space Economy. R.Econ.Statist., 33,1951,318-328
[191] KOOPMANS (T.C.) and REITER (S.): A Model of Transportation. Chap. 14 in Activity Analysis of Production and Allocation. Cowles Comm.Monograph 13, T.C.Koopmans ed.,New-York, John Wiley and Sons, 1951, 222-259.
[192] MEYER-LINDEMANN (H.U.): Typologie der Theorien des Industriestandortes. Veroffentlichungen der Akademie fur Raumforschung und Landesplanung, 21, Walter Dorn Verlag, Bremen-Horn, 1951.
[193] MIKSCH (L.):Zur Theorie des raumlichen Gleichgewichts. Weltwirts. Archiv. 66, 1951,5-47.
[194] MULLER (G.):Raumforschung und Umsiedlung.Bonn, 1951. Institut fur Raumforschung .
[195] PREDOHL (A.): Von der Standortslehre zur Raumwirtschaftslehre. Jb.Sozialwissens, 2,1951, 94-114.

[196] PREDOHL (A.): Le plein emploi dans la perspective spatiale. Econ. Appl., 4,1951,369-391.
[197] STAVENHAGEN (G.): Geschichte der Wirtschaftstheorie. Chap.13. Die Raumwirtschaftslehre, 210-232, Gottingen, 1951.
[198] ULLMAN (E.L.) and ISARD (W.): Toward a More Analytical Economic Geography: the Analysis of Flow Phenomena. Report n.1, Harvard University on Contract to Office of Naval Research. June 1951.
[199] ATKINS (R.M.):A Program for Locating the New Plant. Harvard Business R., 30, 1952,113-121.
[200] BECKMANN (M.J.): Bemerkungen zum Verkehrsgesetz von Lardner. Weltwirts.Archiv., 69,1952,199-213.
[201] BECKMANN (M.J.): A Continuous Model of Transportation. Econometrica 20, 1952,643-660.
[202] DEWEY (R.L.): Criteria for the Establishment of an Optimum Transportation System. Amer.Econ.R. May 1952,644-653.
[203] DUNN (E.S.): The Equilibrium of Land Use Patterns in Agriculture. Doctoral Dissertation, Harvard University Library,1952.
[204] GREENHUT (M.L.): Integrating the Leading Theories of Plant Location. The Southern Economic Journal, 18,1952,526-538.
[205] GREENHUT (M.L.): The Size and Shape of the Market Area of a Firm. The Southern Economic Journal. 19,1952,37-50.
[206] ISARD (W.) and WHITNEY (V.):Atomic Power and Regional Development. Bulletin of Atomic Scientists. 8, 1952,119-124.
[207] ISARD (W.): A General Location Principle of an Optimum Space-Economy. Econometrica. 20,1952,406-430.
[208] ISARD (W.): Current Developments in Regional Analysis. Weltwirts. Archiv. 69,1952,81-90.
[209] LEONTIEF (W.) and ISARD (W.): The Extension of Input-Output Techniques to Interregional Analysis. Part.2 in Studies in the Structure of the American Economy, New-York, Oxford University Press, 1952.
[210] MOSES (L.N.): Regional Input-Output: a Method of Analyzing Regional Interdependence. Ph.D.Dissertation Harvard University. 1952.
[211] SAMUELSON (P.A.): Spatial Price Equilibrium and Linear Programming. Amer.Econ.R. ,June 1952, 283-303.
[212] ALLEN (R.L.): Spatial Patterns and Agglomerations in Location Analysis. Ph.D.Dissertation, Harvard University, 1953.
[213] BECKMANN (M.J.): The Partial Equilibrium of a Continuous Space Market. Weltwirts.Archiv. 71,1953,73-87.
[214] BOUSTEDT (O.): Die Stadtregion. Allgemeines Statistiches Archiv. 1953,13-26.
[215] DITTRICH (E.):Versuch eines Systems der Raumordnung. Bad-Godesberg, Institut fur Raumforschung Bonn, Vertrage 4. 1953.
[216] HARVARD UNIVERSITY: A Revised System for Interregional Analysis. Section G of "Report on Research for 1953" Harvard Economic Research Staff. Dec.19534,91-103.
[217] ISARD (W.):Some Emerging Concepts and Techniques for Regional Analysis. Zeitschrift fur die Gesamte Staatswissenschaft. 109,1953,240-250.
[218] ISARD (W.): Some Remarks in the Marginal Rate of Substitution Between Distance-Inputs and Location Theory. Metroeconomica. 5,1953,11-21.
[219] ISARD (W.):Regional Commodity Balances and Interregional Commodity Flows. Amer.Econ.R., May 1953, 167-180.
[220] ISARD (W.), KAVESH (R.A.),and KUENNE (R.E.): The Economic Base and Structure of the Urban Metropolitan Region. Harvard Economic Research Project. 1953.

[221] KUENNE (R.E.): The Interregional Input-Output Model as a Derivative of a Walrasian Multiple Point System. Regional Section of the Econometrica Society Meeting. Dec.1953.
[222] LEONTIEF (W.W.): Interregional Theory. Chap.4 in Studies in the Structure of the American Economy. New-York,Oxford University Press, 1953,93-115.
[223] POPESCU (O.): Espacio y Economia. Progreso de la Teoria Economica y su importancia para la comprension de las relaciones economicas en las regiones perifericas. Eva Peron.Bahia-Blanca, 1953.
[224] RYAN (J.M.): The Use of Regional Interindustry Analysis in Employment Estimation. Ph.D.Thesis University of North Carolina. 1953.
[225] VINING (R.): Delimitation of Economic Areas: Statistical Conceptions in the Study of the Spatial Structure of an Economic System. Journal of the American Statistical Association. 48,1953,44-64.
[226] VINING (R.): The Study of the Spatial Structure of an Economic System, in Interregional Analysis and Regional Development. Amer.Econ.R. May 1953,p.167ff.
[227] ULLMAN (E.L.): The Basic-Service Ratio and the Areal Support of Cities. Proceedings of the Western Committee on Regional Economic Analysis. Berkeley, Social Science Research Council,1953,110-123.
[228] BONAVIA (M.): The Economics of Transport. London,Cambridge Economic Handbooks,9,1954.
[229] DUNN (E.S.): The Location of Agricultural Production. Gainesville, University of Florida Press, 1954.
[230] ISARD (W.) and PECK (M.J.): Location Theory and International and Interregional Trade Theory. Quart.J.Econ. 68,1954,97-114.
[231] ISARD (W.):Location Theory and Trade Theory: A Short-Run Analysis. Quart.J.Econ. 68,1954,305-320.
[232] LESOURNE (M.J.): Les facteurs economiques de la localisation des entreprises. Nouvelle Revue de l'Economie Contemporaine, Aout-Sept.1954,9-19.
[233] SPENGLER (J.J.):Richard Cantillon: First of the Moderns. J.Polit. Econ., 62,1954,281-295,406-424.
[234] BECKMANN (M.): Some Reflections on Losch's Theory of Location. Papers R.S.A., 1,1955,N.1-N.9.
[235] BECKMANN (M.) and MARSCHAK (T.): An Activity Analysis Approach to Location Theory. Kyklos, 8,1955,125-141.
[236] BECKMANN (M.): The Economics of Location. Kyklos, 8,1955,416-421.
[237] CHRISTALLER (W.):Beitrage zu einer Geographie des Fremdenverkehrs. Erdkunde, 9,1955,1-19.
[238] GOLDNER (W.): Spatial and Locational Aspects of Metropolitan Labor Markets. Amer.Econ.R. 45,1955,113-128.
[239] GREENHUT (M.L.): A General Theory of Plant Location. Metroeconomica. 7,1955.
[240] MEINHOLD (H.):Zahlungsbilanztheorie und Standortstheorie. Jb.Nat. Oekon.Statist. 167,1955,414-423.
[241] MELAMID (A.): Some Applications of Thunen's Model in Regional Analysis of Economic Growth. Papers R.S.A., 1,1955,L.1-L.5.
[242] MOSES (L.N.): The Stability of Interregional Trading Patterns and Input-Output Analysis. Amer.Econ.R., 45,1955,803-832.
[243] NORTH (D.C.): Location Theory and Regional Economic Growth. J.Polit.Econ. June 1955,243-258.
[244] PONSARD (C.): Economie et Espace. Essai d'integration du facteur spatial dans l'analyse economique. Paris,S.E.D.E.S.,1955
[245] POPESCU (O.):La Region Economica. Economica, 3 and 4,1955,p.399ff.

[246] RILEY (V.) and ALLEN (R.L.): Interindustry Economic Studies. A Comprehensive Bibliography on Interindustry Research. Part.8, Regional Analysis. Operations Research Office. The Johns Hopkins University. 1955,177-187.
[247] VALAVANIS (S.): Losch on Location. A Review Article. Amer.Econ.R. Sept.1955,637-644.
[248] VINING (R.): A Description of Certain Spatial Aspects of an Economic System. Economic Development and Cultural Change, 3, 1955,147-198.

FROM 1956 TO 1960

[249] ANDREWS (R.B.): Mechanics of the Urban Economic Base: The Base Concept and the Planning Process. Land Economics. 32, 1956,69-84.
[250] BECKMANN (M.), Mc.GUIRE (C.B.),and WINSTEN (C.B.): Studies in the Economics of Transportation. Cowles Commission for Research in Economics.Yale University Press, New-Haven, 1956.
[251] CARROTHERS (G.A.P.): An Historical Review of the Gravity and Potential Concepts of Human Interaction. Journal of the American Institute of Planners, 22, 1956,94-102.
[252] DUNN (E.S.): The Market Potential Concept and the Analysis of Location. Papers R.S.A., 2,1956,183-194.
[253] FRIEDMANN (J.R.P.): The Concept of a Planning Region, Land Economics. 32,1956,1-13.
[254] FRIEDMANN (J.R.P.): Locational Aspects of Economic Development, Land Economics. 32,1956.
[255] GREENHUT (M.L.): Plant Location in Theory and in Practice. The Economics of Space. Chapel Hill. The University of North-Carolina Press, 1956.
[256] ISARD (W.): Regional Science, the Concept of Region and Regional Structure. Papers R.S.A., 2,1956,13-26.
[257] ISARD (W.): Location and Space-Economy. A General Theory Relating to Industrial Location, Market Areas,Land Use, Trade,and Urban Structure, 1st ed. The Technology Press of Massachusetts Institute of Technology and John Wiley & Sons Inc., New-York, 1956, 2nd ed. Ithaca, Cornell University, Programs in Urban and Regional Studies, 1979. Japanese trans.,1964.
[258] PIATIER (A.): L'attraction commerciale des villes: Un nouveau type d'etude realise en France. French.trans.from Studi di Mercato, July 1956.
[259] PONSARD (C.): Note sur Losch et l'analyse de l'espace economique. Revue d'Economie Politique, 1,1956,75-85.
[260] PONSARD (C.):Note sur la localisation de la firme. Revue Economique, 1,1956,101-116.
[261] RULLIERE (G.): Localisations et Rythmes de l'Activite Agricole. Essai d'analyse economique de la notion de structure agricole. Coll.Etudes et Memoires, 31, Centre d'Etudes Economiques, Paris, Librairie Armand Colin, 1956.
[262] STAMER (H.): Standort und Intensitat. Ein Beitrag zur Standortstheorie des Landbaus. Weltwirts.Archiv. 77, 1956,277-292.
[263] STEINER (R.L.): Urban and Inter-Urban Economic Equilibrium. Land Economics. 32,1956,167-174.
[264] STOLPER (W.F.): Standorttheorie und Theorie des internationalen Handels. Zeitschrift fur die Gesamte Staatswissenschaft, 112,1956,193-217.

[265] STOLPER (W.F.): Teoria della localizzazione e teoria del Commercio Internazionale. Economia Internazionale. 9,1956,459-488.
[266] TIEBOUT (C.M.):A Pure Theory of Local Expenditures. J.Polit.Econ. 64,1956,416-424.
[267] TRIAS-FARGAS (R.): El espacio en el analisis economico. Moneda y Credito, 57,1956,3-29.
[268] BECKMANN (M.J.): International and Interpersonal Division of Labor. Weltwirts.Archiv, 78,1957,67-71.
[269] FLORES (E.): La Economia del Espacio o la Teoria de la Localizacion de la Actividad Economica. Investigacion Economica, 17,1957,331-371.
[270] GREENHUT (M.L.): Games, Capitalism and General Location Theory. The Manchester School of Economic and Social Studies, 25,1957,61-88.
[271] KOOPMANS (T.C.)and BECKMANN (M.):Assignment Problems and the Location of Economic Activities. Econometrica, 25,1957,53-76.
[272] ORR (E.W.):A Synthesis of Theories of Location, of Transport Rates,and of Spatial Price Equilibrium. Papers R.S.A., 3,1957,61-73.
[273] TRIAS-FARGAS (R.): El Concepto Economico de Region: Instrumento Imprescindible del Examen Espacial Empirico. Moneda y Credito, 60,1957,23-46.
[274] WARNTZ (W.): Geography of Prices and Spatial Interaction. Papers R.S.A., 3,1957,118-129.
[275] BECKMANN (M.): City Hierarchies and the Distribution of City Size. Economic Development and Cultural Change, 6,1958,243-248.
[276] BERRY (B.J.L.) and GARRISON (W.L.): Recent Developments of Central Place Theory. Papers R.S.A., 4,1958,107-120.
[277] BERRY (B.J.L.) and GARRISON (W.L.):The Functional Bases of the Central Place Hierarchy. Economic Geography, 34,1958,145-154.
[278] CARROTHERS (G.A.P.): Population Projection by Means of Income Potential Models. Papers R.S.A., 4,1958,121-152.
[279] GOLDMAN (T.A.): Efficient Transportation and Industrial Location. Papers R.S.A., 4,1958,91-106.
[280] ISARD (W.): Interregional Linear Programming. J.R.S., 1958,1,1-59.
[281] LEFEBER (L.): Allocation in Space : Production, Transport and Industrial Location. Amsterdam, North-Holland Publishing Company, 1958.
[282] LEFEBER (L.): General Equilibrium Analysis of Production, Transportation and the Choice of Industrial Location. Papers R.S.A., 4,1958,77-86.
[283] MIEHLE (W.): Link-Length Minimization in Networks. Operations Research, 6,1958,232-243.
[284] MOSES (L.): Location and the Theory of Production. Quart.J.Econ. 72,1958,259-272.
[285] STEVENS (B.H.):An Interregional Linear Programming Model. J.R.S., 1958,1,60-98.
[286] STEWART (C.): The Size and Spacing of Cities. Geographical Review, 48,1958,222-245.
[287] CHAMBRE(H.): L'amenagement du territoire en U.R.S.S. Introduction a l'etude des regions economiques sovietiques. Paris, La Haye, Mouton, 1959.
[288] GROTEWOLD (A.): Von Thunen in Retrospect. Economic Geography, 35, 1959,346-355.
[289] ISARD (W.) and BRAMHALL (F.): Regional Employment and Population Forecasts Via Relative Income Potential Models. Papers R.S.A., 5,1959,25-50.

[290] ISARD (W.) and SCHOOLER (E.W.):Industrial Complex Analysis, Agglomeration Economies, and Regional Development. J.R.S., 1959,1,19-33.
[291] SCHNEIDER (M.): Gravity Models and Trip Distribution Theory. Papers R.S.A., 5,1959,51-56.
[292] SIRKIN (G.): The Theory of the Regional Economic Base. R.Econ. Statist., 41,1959,426-429.
[293] WARNTZ (W.): Toward a Geography of Price: A Study in Geoeconometrics. Philadelphia, University of Pennsylvania Press,1959.
[294] CHRISTALLER (W.): Die Hierarchie der Stadte. Lund Symposium on Problems of Urban Geography, 19th International Geographical Congress. 1960.
[295] HERBERT (J.D.) and STEVENS (B.H.): A Model for the Distribution of Residential Activity in Urban Areas. J.R.S., 2,1960,21-36.
[296] HOLDREN (R.R.): The Structure of a Retail Market and the Market Behavior of Retail Units. Englewoods Cliffs, Prentice-Hall, 1960.
[297] ISARD (W.): Methods of Regional Analysis: An Introduction to Regional Science. New-York,London, The Technology Press of the Massachusetts Institute of Technology and John Wiley and Sons, Inc., 1960.
[298] MOSES (L.N.): A General Equilibrium Model of Production, Interregional Trade , and Location of Industry. R.Econ. Statist. 42,1960,373-399.
[299] STOUFFER (S.A.): Intervening Opportunities and Competing Migrants. J.R.S. 2,1960,1-26.
[300] ULLMAN (E.L.) and DACEY (M.F.):The Minimum Requirements Approach to the Urban Economic Base. Papers R.S.A., 6,1960,175-194.

FROM 1961 TO 1965

[301] MUTH (R.F.): Economic Change and Rural-Urban Land Conversions. Econometrica, 24,1961,1-23.
[302] STEVENS (B.H.):An Application of Game Theory To a Problem in Location Strategy. Papers R.S.A., 7,1961,143-157.
[303] THOMAS(E.N.):Toward an Expanded Central Place Model. Geographical Review, 51,1961,400-411.
[304] TIEBOUT (C.M.): An Economic Theory of Fiscal Decentralization. in NBER, Public Finances: Needs, Sources and Utilization, Princeton University Press, 1961.
[305] TINBERGEN (J.):The Spatial Dispersion of Production: A Hypothesis. Schweizerisch Zeitschrift fur Volkswirtschaft und Statistik, 97,1961,412-419.
[306] WINGO (L.J.): Transportation and Urban Land. Resources for the Future, Inc. Baltimore, John Hopkins Press, 1961.
[307] WINGO (L.J.):An Economic Model of the Utilization of Urban Land for Residential Purposes. Papers R.S.A., 7,1961,191-205.

[308] BOVENTER (E.von): Die Struktur der Landschaft; Versuch einer Synthese und Weiterentwicklung der Modelle J.H.von Thunens, W.Christallers und A.Loschs. Schriften des Vereins fur Socialpolitik, N.F. 27,1962,77-133.
[309] BOVENTER (E.von): Towards a United Theory of Spatial Economic Structure. Papers R.S.A., 10,1962,163-187.
[310] BOVENTER (E.von): Theorie des raumlichen Gleichgewichts. Tubingen, J.C.B.Mohr (Paul Siebeck),1962. French trans. by DALOZ (J.P.): Theorie de l'equilibre en economie spatiale. Paris, Gauthiers-Villars, 1966.
[311] DACEY (M.F.): Analysis of Central Place and Point Pattern by Nearest Neighbor Method. Lund Studies in Geography, Series B,24,1962,55-75.
[312] HUFF (D.L.): Determination of Intra-Urban Retail Trade Areas. Los Angeles, Real Estate Research Programm, The University of California, 1962.
[313] KUHN (H.W.) and KUENNE (R.E.):An Efficient Algorithm for the Numerical Solution of the Generalized Weber Problem in Spatial Economics. J.R.S., 4,1962,21-33.
[314] TIEBOUT (C.M.): The Community Economic Base Study. New-York, Committee for Economic Development, 1962.
[315] CHRISTALLER (W.): Wandlungen des Fremdenverkehrs und der Bergstrasse, im Odenwald und im Neckartal. Geographische Rundschau, 15,1963,216-222.
[316] COOPER (L.): Location - Allocation Problems. Operations Research, 11,1963,331-343.
[317] GREENHUT (M.L.): Microeconomics and the Space-Economy. Chicago, Scott-Foresman, 1963.
[318] HUFF (D.L.): A Probabilistic Analysis of Shopping Center Trade Areas. Land Economics, Feb.1963,81-90.
[319] JACOT (S.P.): Strategie et Concurrence: de l'application de la theorie des jeux a l'analyse de la concurrence spatiale. Paris, SEDES,1963.
[320] ROBINE (M.): Note sur l'estimation statistique des parametres de la loi de Reilly. Travaux et Documents de l'Institut d'Administration des Entreprises de Bordeaux, 1963.
[321] STOLLSTEIMER (J.F.): A Working Model for Plant Numbers and Location. Journal of Farm Economics, 43,1963,631-645.
[322] TOBLER (W.R.): Geographic Area and Map Projection. Geographical Review, 52,1963,59-78.
[323] WINGO (L.J.)ed: Cities and Space: the Future Use of Urban Space. Resources for the Future, inc.John Hopkins Press. Baltimore, 1963.
[324] ALONSO (W.): Location and Land-Use: Toward a General Theory of Land Rent. Cambridge,Mass.,Harvard University Press,1964.
[325] BALINSKI (M.L.): On Finding Integer Solutions to Linear Programs. Mathematica, Princeton, 1964.
[326] BOUCHARD (R.J.) and PYERS (C.E.):Use of Gravity Models for Describing Urban Travel. Highway Research Record, 88,1964,1-43.

[327] CHRISTALLER (W.): Some Considerations of Tourism Location in Europe: The Peripheral Regions - Underdeveloped Countries - Recreation Areas. Papers R.S.A., 12,1964,95-105.
[328] COOPER (L.): Heuristic Methods for Location - Allocation Problems. Siam Review, 6,1964,37-52.
[329] DACEY (M.F.): Modified Probability Law for Point Pattern More Regular than Random. Annals of the Association of American Geographers, 54,1964,559-565.
[330] DACEY (M.F.): A Family of Density Functions of Losch's Measurements on Town Distribution. Professional Geographer, 16,1964,5-7.
[331] DACEY (M.F.): A Note on Some Number Properties of a Hexagonal Hierarchical Plane Lattice. J.R.S., 5,1964,63-67.
[332] HAKIMI (S.L.): Optimum Location of Switching Centers and the Absolute Centers and Medians of a Graph. Operations Research, 12,1964,450-459.
[333] HARRIS (B.): A Note on the Probability of Interaction at a Distance. J.R.S., 5,1964,31-35.
[334] LOWRY (I.S.): A Model of Metropolis. Santa Monica, California, The Rand Corporation, Memorandum R4,4035,RL, 1964.
[335] MANNE (A.S.):Plant Location Under Economies of Scale - Decentralization and Computation. Management Science, 11,1964,213-235.
[336] MILLS (E.S.) and LAV (M.R.): A Model of Market Areas with Free Entry. J.Polit.Econ., 72,1964,278-288.
[337] SWEET (F.H.): An Error Parameter for the Reilly-Converse Law of Retail Gravitation. J.R.S., 5,1964,69-72.
[338] TAKAYAMA (T.) and JUDGE (G.G.):Equilibrium Among Spatially Separated Markets: A Reformulation. Econometrica, 32,1964.
[339] TAKAYAMA (T.) and JUDGE (G.G.): An Intertemporal Price Equilibrium Model. Journal of Farm Economics, 46,2,1964.
[340] TAKAYAMA (T.) and JUDGE (G.G.): Spatial Equilibrium and Quadratic Programming. Journal of Farm Economics, 46,1,1964.
[341] TINBERGEN (J.): Sur un modele de la dispersion geographique de l'activite economique. Revue d'Economie Politique, 1, 1964,30-44.
[342] VICKERY (W.S.): Microstatics. New-York, Harcourt, Brace and World, 1964.
[343] BEEN (R.O.): A Reconstruction of the Classical Theory of Location. University of California, Berkeley, Unpublished Ph.D. Dissertation, 1965.
[344] BERRY (B.J.L.)and PRED(A.): Central Place Studies. A Bibliography of Theory and Applications. Philadelphia,Regional Science Research Institute, Bibliography Series,1,1961. Supplement through 1964 by BARNUM (H.G.),KASPERSON (R.), and KIUCHI (S.),1965.
[345] BOS (H.C.): Spatial Dispersion of Economic Activity. Rotterdam University Press, 1965.
[346] DACEY (M.F.): Order Distance in an Inhomogeneous Random Point Pattern. Canadian Geographer, 9,1965,144-153.
[347] DACEY (M.F.): The Geometry of Central Place Theory. Geografiska Annaler, 47B, 1965,111-124.
[348] DACEY (M.F.): An Interesting Number Property in Central Place Theory. Professional Geographer, 17,1965,32-33.
[349] DEVLETOGLOU (N.E.): A Dissenting View of Duopoly and Spatial Competition. Economica, 32,1965,140-160.
[350] KUHN (H.W.): Locational Problems and Mathematical Programming.in Applications of Mathematics to Economics, Budapest, Publishing House of the Hungrarian Academy of Sciences, 1965,235-248.

[351] LACHENE (R.): Contribution a l'analyse de l'espace economique. Metra, 6,1965,5-146.
[352] MATHERON (G.): Les variables regionalisees et leur estimation. Paris, Masson,1965.
[353] MORAN (P.): L'analyse spatiale en Sciences economiques. Paris, Cujas, 1965.
[354] OLSSON (G.): Distance and Human Interaction: A Review and Bibliography, Philadelphia, Regional Science Research Institute Bibliography Series, 2,1965.
[355] WARNTZ (W.): Macrogeography and Income Fronts. Philadelphia, The Regional Science Research Institute, Monograph Series,3, 1965.
[356] WITZGALL (C.): Optimal Location of a Central Facility: Mathematical Models and Concepts. National Bureau of Standards, Report 8388, 1965.

FROM 1966 TO 1970

[357] BECKMANN (M.): Location Theory. Chicago, Rand-McNally,1966.
[358] DACEY (M.F.): A Compound Probability Law for a Pattern More Dispersed than Random and with Areal Inhomogeneity. Economic Geography, 42,1966,172-179.
[359] DACEY (M.F.): Population of Places in a Central Place Hierarchy. J.R.S., 6,1966,27-33.
[360] DACEY (M.F.): A Probability Model for Central Place Locations. Annals of the Association of American Geographers, 56, 1966,550-568.
[361] NEFT (D.S.): Statistical Analysis for Areal Distributions. Philadelphia, Regional Science Research Institute, Monograph Series, 2,1966.
[362] PONSARD (C.):Une application de la theorie des graphes a l'analyse de l'espace economique:un modele de localisation optimale de l'unite de production dans une structure de concurrence.Techniques Economiques Modernes, 4,1966,1-21.
[363] STEVENS (B.H.) and RYDELL (C.P.):Spatial Demand Theory and Monopoly Price Policy. Papers R.S.A., 17,1966,195-204.
[364] ALONSO (W.): A Reformulation of Classical Location Theory and Its Relation to Rent Theory. Papers R.S.A., 19,1967,23-44.
[365] BERRY (B.J.L.): Geography of Market Centers and Retail Distribution. Englewood Cliffs,N.J., Prentice-Hall,Inc.Foundations of Economic Geography Series, 1967.
[366] BOVENTER (E.von): Land Values and Spatial Structure: Agricultural, Urban and Tourist Location Theories. Papers R.S.A., 18, 1967,231-242.
[367] CLARK (C.): Von Thunen's Isolated State. Oxford Economic Papers, 19,1967,370-377.
[368] COOPER (L.): Solutions of Generalized Locational Equilibrium Models. J.R.S., 7,1967,1-18.
[369] CURRY (L.): Central Places in the Random Spatial Economy. J.R.S., 7,1967,217-238.
[370] DEVLETOGLOU (N.E.) and DEMETRIOU (P.A.): Choice and Threshold: A Further Experiment in Spatial Duopoly. Economica, 1967, 251-271.
[371] DZIEWONSKI (K.): The Concept of the Urban Economic Base: Overlooked Aspects. Papers R.S.A., 18,1967,139-145.
[372] FAYETTE (J.R.): Contribution a la theorie micro-economique de l'optimum spatial. University of Lyon, Unpublished Doctoral Dissertation, 1967.

[373] GOKHMAN (V.M.) and LIPETS (Y.G.): Some Trends of Soviet Regional Studies. Papers R.S.A., 18,1967,223-229.
[374] HUDSON (J.C.): An Algebraic Relation Between the Losch and Christaller Central Place Networks. Professional Geographer, 19,1967,133-135.
[375] ISARD (W.): Game Theory, Location Theory and Industrial Agglomeration. Papers R.S.A., 18,1967,1-11.
[376] ISARD (W.) and SMITH (T.E.): Location Games: with Applications to Classic Location Problems. Papers R.S.A., 19,1967,45-80.
[377] LEVY (J.): An Extended Theorem for Location on a Network. Operational Research Quarterly, 18,1967,433-443.
[378] OLSSON (G.): Central Place System, Spatial Interaction, and Stochastic Processes. Papers R.S.A., 18,1967,13-45.
[379] STEVENS (B.H.) and BRACKETT (C.A.): Industrial Location. A Review and Annotated Bibliography of Theoretical, Empirical and Case Studies. Philadelphia, Regional Science Research Institute. Bibliography Series, 3, 1967.
[380] VERGIN (R.C.) and ROGERS (J.D.): An Algorithm and Computational Procedure for Locating Economic Facilities. Management Science, 13,B240-B254, 1967.
[381] BECKMANN (M.): Location Theory. New-York, Random House, 1968.
[382] BROWN (L.A.): Diffusion Processes and Location. A Conceptual Framework and Bibliography. Philadelphia, Regional Science Research Institute, Bibliography Series, 4,1968.
[383] COLE (J.P.)and KING (C.A.M.): Quantitative Geography. Techniques and Theories in Geography. London, John Wiley and Sons,1968.
[384] COOPER (L.): An Extension of the Generalized Weber Problem. J.R.S. 8,1968,181-197.
[385] DACEY (M.F.) and SEN (A.):Complete Characterization of the Central Place Hexagonal Lattice. J.R.S., 8,1968,209-213.
[386] GAMBINI (R.), HUFF (D.L.)and JENKS (G.F.): Geometric Properties of Market Areas. Papers R.S.A., 20,1968,85-92.
[387] ISARD (W.) and SMITH (T.E.):On Social Decision Procedures for Conflict Situations. Papers Peace Research Society, 8,1968, 1-29.
[388] ISARD (W.) and SMITH (T.E.): Coalition Location Games: Paper 3. Papers R.S.A., 20,1968,95-107.
[389] JONAS (P.): Spatial Competition in a Two Dimensional Market. The Annals of Regional Science, 1968,12-29.
[390] MEDVEDKOV (Y.V.): An Application of Topology in Central Place Analysis. Papers R.S.A., 20,1968,77-84.
[391] NOURSE (H.O.): Regional Economics. New-York, McGraw-Hill Book Company. 1968.
[392] PONSARD (C.): Les modeles de hierarchisation et la pseudo-fonction de Grundy. Rivista Internazionale di Scienze Economiche e Commerciali, 2, 1968,122-131.
[393] SAKASHITA (N.): Production Function, Demand Function and Location Theory of the Firm. Papers R.S.A., 20,1968,109-122.
[394] SEYMOUR (D.R.): The Polygon of Forces and the Weber Problem. J.R.S. 8,1968,243-246.
[395] STEVENS (B.H.): Location Theory and Planning Models: The Von Thunen Case. Papers R.S.A., 21,1968,19-34.
[396] TAKAYAMA (T.) and JUDGE (G.G.): Alternative Spatial Equilibrium Models. J.R.S., 10,1968,1-12.
[397] TEITZ (M.B.): Locational Strategies for Competitive Systems. J.R.S., 8,1968,135-148.
[398] TEITZ (M.B.): Towards a Theory of Urban Facility Location. Papers R.S.A., 21, 1968, 35-52.

[399] TINBERGEN (J.): The Hierarchy Model of the Size Distribution of Centers. Papers R.S.A., 20,1968,65-68.
[400] BOVENTER (E.von): Walter Christaller's Central Places and Peripheral Areas: The Central Place Theory in Retrospect. J.R.S. 9, 1969,117-124.
[401] CURRY (L.): A "Classical" Approach to Central Place Dynamics. Geographical Analysis, 1969,1,272-282.
[402] DAY (R.H.) and TINNEY (E.H.): A Dynamic von Thunen Model. Geographical Analysis, 1,1969,137-151.
[403] DICKINSON (H.O.):Von Thunen's Economics. Econ J., 79,1969,894-902.
[404] DOCKES (P.): L'espace dans la pensee economique du 16e au 18e siecle. Paris, Flammarion, 1969. Italian trans.by STEFANIS (M.de): Lo spazio nel pensiero economico dal 16 al 18 secolo. Milano, Feltrinelli Editore, 1971.
[405] FANO (P.L.): Organization, City Size Distributions and Central Places. Papers R.S.A., 22,1969,29-38.
[406] HYMAN (G.M.): The Calibration of Trip Distribution Models. Environment and Planning, A,1,1969,105-112.
[407] ISARD (W.)and al: General Theory: Social, Political, Economic,and Regional with Particular Reference to Decision-Making Analysis. Cambridge, M.I.T.Press,1969.
[408] MATHIESON (R.S.): The Soviet Contribution to Regional Science: A Review Article. J.R.S., 9,1969,125-140.
[409] MENNES (L.B.M.),TINBERGEN (J.)and WAARDENBURG (J.G.): The Element of Space in Development Planning. Amsterdam,London,North-Holland Publishing Company, 1969.
[410] MUTH (R.F.): Cities and Housing: The Spatial Pattern of Urban Residential Land Use. Chicago, University of Chicago Press,1969.
[411] NIEDERCORN (J.H.)and BECHDOLT (B.V.):An Economic Derivation of the Law of Spatial Interaction. J.R.S., 9,1969,273-282.
[412] OORT (C.J.): La theorie marginaliste et les prix de transport. Rotterdam,Fondation Verkeerswetmschappelijk Contrum,1969.
[413] PARR (J.B.): City Hierarchies and the Distribution of City Size: A Reconsideration of Beckmann's Contribution. J.R.S. 9,1969,239-254.
[414] PONSARD (C.): Un modele topologique d'equilibre economique interregional. Paris, Dunod, 1969.
[415] PONSARD (C.): Theorie de la base economique et croissance urbaine. in Developpement urbain et Analyse economique. Paris, Cujas, 1969, 125-148.
[416] RICHARDSON (H.W.): Regional Economics: Location Theory, Urban Structure, Regional Change. New-York, Frederick A.Praeger,1969.
[417] RICHARDSON (H.W.): Elements of Regional Economics. Baltimore, Penguin, 1969.
[418] SCHARLIG (A.): Localisation optimale et Theorie des graphes. Geneve, Cahiers V.Pareto, 19, Librairie Droz, 1969. Spanish trans.by ARGAMENTERIA GARCIA (R.): Localizacion Optima y Teoria de Grafos. Madrid, Instituto de Estudios Politicos, 1973.
[419] SIEBERT (H.): Regional Economic Growth: Theory and Policy. Scranton,Pennsylvania, International Textbook Company,1969.
[420] TELSER (L.G.): On the Regulation of Industry. J.Polit.Econ., 77, 1969, 937-952.
[421] TERMOTE (M.): Migration et equilibre economique spatial. Louvain, Faculte des Sciences Economiques, Sociales et Politiques, 54,1969.

[422] BECKMANN (M.J.) and McPHERSON (J.C.): City Size Distribution in a Central Place Hierarchy: An Alternative Approach. J.R.S., 10,1970,25-33.
[423] BRESSLER (R.G.Jr.) and KING (R.A.): Markets, Prices, and Interregional Trade. New-York, John Wiley and Sons, 1970.
[424] DEAN (R.D.), LEAHY (W.H.), and McKEE (D.L.)eds: Spatial Economic Theory. New-York, Free Press, 1970.
[425] DENIKE (K.G.) and PARR (J.B.): Production in Space, Spatial Competition, and Restricted Entry. J.R.S., 10,1970,49-63.
[426] GREENHUT (M.L.): A Theory of the Firm in Economic Space, New-York, Meredith Corporation, 1970.
[427] HOOVER (E.M.): Transport Costs and the Spacing of Central Places. Papers R.S.A., 25,1970,255-274.
[428] LEAHY (W.H.), McKEE (D.L.), and DEAN (R.D.) eds: Urban Economics: Theory, Development, and Planning, New-York, Free Press, 1970.
[429] LOVELL (M.): Product Differentiation and Market Structure. Western Economic Journal, 8,1970,120-143.
[430] MATHUR (V.K.): An Economic Derivation of the Gravity Law of Spatial Interaction: a Comment. J.R.S., 10,1970,403-405.
[431] MILLS (E.S.): The Efficiency of Spatial Competition. Papers R.S.A. 25,1970,71-82.
[432] NIEDERCORN (J.H.)and BECHDOLT (B.V.):An Economic Derivation of the Gravity Law of Spatial Interaction: Reply. J.R.S., 10, 1970, 407-410.
[433] PARR (J.B.): Models of City Size in an Urban System. Papers R.S.A., 25,1970,221-253.
[434] SERCK-HANSSEN (J.): Optimal Patterns of Location. Amsterdam, North-Holland, 1970.
[435] SILBERBERG (E.): A Theory of Spatially Separated Markets. International Economic Review, 11,1970.
[436] SMOLENSKY (E.), BURTON (R.), and TIDEMAN (N.): The Efficient Provision of a Local Non-Private Good. Geographical Analysis, 2,1970, 330-342
[437] TREUNER (P.): An Infrastructure Cost Model for a System of Central Places. Papers R.S.A., 24,1970,85-101.
[438] WILSON (A.G.): Entropy in Urban and Regional Modelling. Monographs in Spatial and Environmental Systems Analysis, London, Pion Limited, 1970.

FROM 1971 TO 1975

[439] ALLEN (W.B.): Developing and Testing of a Behavioral Modal Split Model. Transportation Studies Center, Philadelphia,University of Pennsylvania, June 1971.
[440] BECKMANN (M.): Equilibrium Versus Optimum: Spacing of Firms and Patterns of Market Areas. Northeast Regional Science Review ,1,1971,1-20.
[441] BECKMANN (M.): Spatial Oligopoly Revisited: An Examination of Some Strategies in Mill Pricing in One and Two Dimensional Markets, Papers R.S.A., 28,1971,37-47.
[442] BECKMANN (M.J.): Market Share, Distance and Potential. R.U.E., 1,1971,3-18.
[443] BECKMANN (M.) and GOLOB (T.F.): A Critique of Entropy and Gravity in Travel Forecasting. Paper,5th International Symposium, on the Theory of Traffic Flow and Transportation, Berkeley, California, June 1971.

[444] BRADFIELD (M.): A Note on Location and the Theory of Production. J.R.S., 11,1971,263-266.
[445] EILON (S.),WATSON-GANDY (C.D.T.),and CHRISTOFIDES (N.): Distribution Management: Mathematical Modelling and Practical Analysis. London, Charles Griffin, 1971.
[446] EVANS (A.W.): The Calibration of Trip Distribution Models with Exponential or Similar Cost Function. Transportation Research, 5,1971,15-38.
[447] GADREAU (M.): Localisation et Oligopole. Coll.IME, 1, Paris, Sirey,1971.
[448] GANNON (C.A.): Fundamental Properties of Loschian Spatial Demand. Environment and Planning, 3,1971,283-306.
[449] GUIGOU (J.L.): On French Location Models for Production Units. R.U.E., 1,1971, 2,107-138, and 3, 289-316.
[450] HARTWICK (J.) and HARTWICK (P.): Duopoly in Space. Canadian Journal of Economics, 4,1971,485-505.
[451] ISARD (W.): Spatial Interaction Analysis: Some Suggestive Thoughts from General Relativity Physics. Papers R.S.A., 27,1971, 17-38.
[452] LONG (W.H.): Demand in Space: Some Neglected Aspects. Papers R.S.A. 27,1971,45-60.
[453] OLSSON (G.): Analogs, Theories and Decision Making: Comments on Walter Isard's Paper. Papers R.S.A., 27,1971,39-43.
[454] PONSARD (C.): Note sur la localisation de la branche d'activite dans une structure de concurrence quasi-parfaite. Revue d'Economie Politique, 6,1971,1028-1031.
[455] RIEGGER (R.)ed: August Losch. In Memoriam. Heidenheim, Verlag der Buchhandlung Meuer, 1971.
[456] ROUGET (B.): Modeles de gravitation et theorie des graphes. Coll. I.M.E.,2, Paris, Sirey, 1971.
[457] SCOTT (A.J.): Location - Allocation Systems. A Review. Geographical Analysis, 3,1971, 95-119.
[458] SCOTT (A.J.): Combinatorial Programming, Spatial Analysis, and Planning. London, Methuen, 1971.
[459] SCOTT (A.J.): Optimal Decision Processes for a Class of Dynamic Locational Problems. Papers R.S.A., 26,1971,25-35.
[460] TAIEB (F.) and DIMEGLIO (P.): Un modele de localisation des surfaces commerciales: Paprica. Urbanisme, 126,1971,10-17.
[461] TAKAYAMA (T.) and JUDGE (G.G.): Spatial and Temporal Price and Allocation Models. Amsterdam, North-Holland, 1971.
[462] ALLEN (W.B.): An Economic Derivation of the Gravity Law of Spatial Interaction: a Comment on the Reply. J.R.S. 12,1972,119-126.
[463] ANDERSSON (A.E.) and MARKSJO: General Equilibrium Models for Allocation in Space Under Interdependency and Increasing Returns to Scale. R.U.E., 2,1972,133-158.
[464] BABUSIAUX (D.): Establishing a Network of Warehouses for Finished Products, Taking Into Account Seasonal Fluctuations in Demand. R.U.E., 1,1972,337-353.
[465] BECKMANN (M.): Von Thunen Revisited. A neo-classical Land Use Model. Swedish Journal of Economics, 74,1972, 1-7.
[466] BECKMANN (M.): Spatial Cournot Oligopoly. Papers R.S.A. 28,1972. 37-47.
[467] BECKMANN (M.J.) and SCHRAMM (G.): The Impact of Scientific and Technical Change on the Location of Economic Activities. R.U.E., 2,1972, 159-174.
[468] BOLLOBAS (B.) and STERN (N.H.): The Optimal Structure of Market Areas. Journal of Economic Theory, 4,1972,174-179.

[469] BOUDEVILLE (J.R.): Amenagement du territoire et polarisation, Paris, Genin, 1972.
[470] COOPER (L.): The Transportation - Location Problem. Operations Research, 20,1972,94-108.
[471] DZIEWONSKI (K.): General Theory of Rank-Size Distributions in Regional Settlement Systems: Reappraisal and Reformulation of the Rank-Size Rule. Papers R.S.A., 29,1972,73-86.
[472] EATON (B.C.): Spatial Competition Revisited. Canadian Journal of Economics, 5,1972,268-277.
[473] FALES (R.L.) and MOSES (L.N.): Land-Use Theory and the Spatial Structure of the Nineteenth - Century City. Papers R.S.A., 28,1972,49-80.
[474] GADREAU (M.): Location and Oligopoly Location of Production Units When Location Decisions Are Interdependent. R.U.E., 2, 1972,175-217.
[475] GANNON (C.A.): Consumer Demand, Conjectural Interdependance and Location Equilibria in Simple Spatial Duopoly. Papers R.S.A., 28,1972,83-107.
[476] GOULD (P.): Pedagogic Review of A.G.Wilson's Entropy in Urban and Regional Modelling. Annals of the Association of American Geographers, 62,1972,689-700.
[477] GREENHUT (M.L.) and OHTA (H.): Monopoly Output Under Alternative Spatial Pricing Techniques. Amer.Econ.R., 62,1972,705-713.
[478] GREENHUT (M.L.), OHTA (H.), and SCHEIDELL (J.): A Model of Market Area Under Discriminatory Pricing. Western Economic Journal, 10, 1972, 402-413.
[479] GUIGOU (J.L.): Theorie economique et transformation de l'espace agricole. 2 vol., Paris, Gauthier-Villars, 1972.
[480] HANSEN (S.): Utility, Accessibility and Entropy in Spatial Modelling. The Swedish Journal of Economics, 74,1972,35-44.
[481] HANSEN (P.) and KAUFMAN (L.): Comparaison d'algorithmes pour le probleme de la localisation des entrepots. in: BRENNAN (J.),ed., Operational Research in Industrial Systems, London, English University Press, 1972,281-294.
[482] HARTWICK (J.M.): The Gravity Hypothesis and Transportation Cost Minimization. R.U.E., 2,1972,297-308.
[483] HEFFLEY (D.R.): The Quadratic Assignment Problem: A Note. Econometrica, 40,1972,1155-1163.
[484] HENDERSON (J.V.): Hierarchy Models of City Size: An Economic Evaluation. J.R.S., 12,1972,435-441.
[485] HURTER (A.P.) and WENDELL (R.E.): Location and Production. A Special Case. J.R.S., 12,1972,243-247.
[486] ISARD (W.): Ecologic-Economic Analysis for Regional Development: Some Initial Explorations with Particular Reference to Recreational Resource Use and Environmental Planning. New-York, Free Press, 1972.
[487] KUENNE (R.E.) and SOLAND (R.M.): Exact and Approximate Solutions to the Multisource Weber Problem. Mathematical Programming, 3,1972,193-209.
[488] MILLS (E.S.): Urban Economics, Glenview, Scott Foresman, 1972.
[489] NIEDERCORN (J.H.) and BECHDOLT (B.V.): An Economic Derivation of the Gravity Law of Spatial Interaction: a Further Reply and a Reformulation. J.R.S., 12, 1972, 127-136.
[490] PONSARD (C.)ed: Graphes de transfert et Analyse economique. Paris, Sirey, Revue d'Economie Politique et Institut de Mathematiques Economiques, 1972.
[491] ROUGET (B.): Graph Theory and Hierarchisation Models. R.U.E., 2, 1972,263-295.

[492] RUSHTON (G.): Map Transformations of Point Patterns: Central Place Patterns in Areas of Variable Population Density. Papers R.S.A., 28,1972,111-129.
[493] SARLY (R.M.):A Model for the Location of Rural Settlement. Papers R.S.A., 29, 1972, 87-104.
[494] STERN (N.H.): The Optimal Size of Market Areas. Journal of Economic Theory, 4,1972, 154-173.
[495] TELLIER (L.N.): The Weber Problem: Solution and Interpretation. Geographical Analysis, 4, 1972, 214-233.
[496] WEBBER (M.J.): Impact of Uncertainty on Location. Cambridge,M.I.T. Press, 1972.
[497] ZOLLER (H.G.) Localisation residentielle: Decision des menages et Developpement suburbain. Bruxelles, Les Editions Vie Ouvriere, 1972.
[498] ANAS (A.): A Dynamic Disequilibrium Model of Residential Location. Environment and Planning, 5,1973,633-647.
[499] BEAVON (K.S.O.): A Procedure for Constructing Losch's Regional System of Markets. South African Journal of Science, 69, 1973,377-379.
[500] BECKMANN (M.J.): The Isolated Region: A Model of Regional Growth. R.U.E., 3,1973,223-231.
[501] BECKMANN (M.J.): Equilibrium Models of Residential Land Use. R.U.E., 3,1973,361-368.
[502] COOPER (L.): N-Dimensional Location Models: An Application to Cluster Analysis. J.R.S., 13,1973, 41-54.
[503] DOKMECI (V.F.): An Optimization Model for a Hierarchical Spatial System. J.R.S., 13, 1973,439-451.
[504] EATON (B.C.): Comment On "Duopoly and Space". Canadian Journal of Economics, 6, 1973,124-217.
[505] EMERSON (D.L.): Optimum Firm Location and the Theory of Production. J.R.S., 13,1973,335-347.
[506] FORSUND (F.R.): Externalities, Environmental Pollution and Allocation in Space: A General Equilibrium Approach. R.U.E. 3,1973,3-32.
[507] GANNON (C.A.): Optimization of Market Share in Spatial Competition. Southern Economic Journal, 40,1973,66-79.
[508] GANNON (C.A.):Central Concentration in Simple Spatial Duopoly: Some Behavioral and Functional Conditions. J.R.S., 13, 1973,357-375.
[509] GOLOB (T.F.), GUSTAFSON (R.L.), and BECKMANN (M.J.): An Economic Utility Theory Approach to Spatial Interaction. Papers R.S.A., 30,1973,159-182.
[510] HARTWICK (J.M.): Losch's Theorem on Hexagonal Market Areas. J.R.S., 13,1973,213-221.
[511] HURIOT (J.M.) and PERREUR (J.): Modeles de localisation et distance rectilineaire. Revue d'Economie Politique, 83,1973,640-662.
[512] HURIOT (J.M.) and PERREUR (J.): On the Weber Problem with Rectangular Distance: A Comment. Management Science, 20,1973, 418-419.
[513] JUDGE (G.G.)and TAKAYAMA (T.)eds: Studies in Economic Planning over Space and Time. New-York, American Elsevier, 1973.
[514] NIJKAMP (P.)and PAELINCK (J.): A Solution Method for Neo-Classical Location Problems. R.U.E., 3, 1973,383-410.
[515] PARR (J.B.): Structure and Size in the Urban System of Losch. Economic Geography, 49,1973,185-212.
[516] ROSENHEAD (J.) and POWELL (G.): The Ice-cream Man Problem. Transpn.Res., 9,1973, 117-121.
[517] SCHARLIG (A.): About the Confusion between the Center of Gravity and Weber's Optimum. R.U.E., 3,1973, 371-382.

[518] SOUFFLET (J.F.): The Warehouse Problem: A Review. R.U.E., 3, 1973,187-216.
[519] TARRANT (J.R.): Comments on the Losch Central Place System. Geographical Analysis, 5,1973,113-121.
[520] WENDELL (R.E.) and HURTER (A.P.): Optimal Locations on a Network. Transportation Science, 7, 1973, 18-33.
[521] WENDELL (R.E.) and HURTER (A.P.): Location Theory, Dominance, and Convexity. Operations Research, 21, 1973,314-320.
[522] WOODWARD (R.S.): The Iso-Outlay Function and Variable Transport Costs. J.R.S., 13,1973, 349-355.
[523] BECKMANN (M.J.): A Theorem on Perfect Competition in Spatial Markets. Papers R.S.A., 33,1974, 3-12.
[524] BOUNON (C.): Spatial Equilibrium of the Sector in Quasi-Perfect Competition. Working Paper, I.M.E., 3, 1974.
[525] COOPER (L.): A Random Locational Equilibrium Problem. J.R.S., 14,1974, 47-54.
[526] DAVIES (O.): Optimal Facility Location in a One-Dimensional Spatial Market. Geographical Analysis, 6,1974,239-264.
[527] DEARING (P.M.) and FRANCIS (R.L.): A Network Flow Solution to a Multifacility Minimax Location Problem Involving Rectilinear Distances. Transportation Science, 8,1974, 126-141.
[528] FLATTERS (F.), HENDERSON (V.), and MIESZKOWSKI (P.): Public Goods, Efficiency and Regional Fiscal Equalization. Journal of Public Economics, 3,1974, 99-112.
[529] HAMILTON (F.E.I.)ed: Spatial Perspectives on Industrial Organization and Decision-Making. London,John Wiley and Sons,1974
[530] HARTWICK (J.M.): Price Sustainability of Location Assignments. Journal of Urban Economics, 1, 1974,147-160.
[531] HURIOT (J.M.): L'economie de l'espace. Concepts et methodes; in: Theorie economique et utilisation de l'espace. Paris, Cujas, Cahiers T.E.M.Espace, 6,1974,5-20.
[532] KATZ (I.N.) and COOPER (L.): An Always-Convergent Numerical Scheme for a Random Locational Equilibrium Problem. SIAM Journal of Numerical Analysis, 11,1974,683-692.
[533] KATZMAN (M.T.): The Von Thunen Paradigm, the Industrial - Urban Hypothesis, and the Spatial Structure of Agriculture. American Journal of Agricultural Economics, 56,1974,683-696.
[534] KHALILI (A.), MATHUR (V.K.), and BODENHORN (D.): Location and the Theory of Production:A Generalization. Journal of Economic Theory, 9,1974,467-475.
[535] KLAASSEN (L.H.):A Shopping Model.Rotterdam, Netherlands Economic Institute, Foundations of Empirical Economic Research,1974.
[536] LANTNER (R.)and THISSE (J.F.): Une revision des conditions d'equilibre de la firme: l'integration du facteur spatial. Revue d'Economie Politique, 1,1974,108-113.
[537] NIEDERCORN (J.H.) and MOOREHEAD (J.D.): The Commodity Flow Gravity Model. A Theoretical Reassessment. R.U.E., 4,1974,68-75.
[538] NIJKAMP (P.)and PAELINCK (J.H.P.): A Dual Interpretation and Generalization of Entropy-Maximization Models in Regional Science. Papers R.S.A., 33,1974,13-31.
[539] OSLEEB (J.P.): The Optimum Size of Plant for the Uniform Delivered Price Manufacturer. Proceedings of the Association of American Geographers, 6,1974,102-105.
[540] PERREUR (J.): Contribution a la theorie de la localisation de l'entreprise. University of Dijon, Unpublished Doctoral Dissertation, 1974.

[541] PERREUR (J.) and THISSE (J.F.): Central Metrics and Optimal Location. J.R.S., 14,1974,411-421.
[542] PERREUR (J.) and THISSE (J.F.): Une application de la metrique circumradiale a l'etude des deplacements urbains. Revue Economique, 25,1974, 298-315.
[543] PERRIN (J.C.): Le developpement regional, Paris, Presses Universitaires de France, 1974.
[544] PONSARD (C.): Une revision de la theorie des aires de marche. Coll.I.M.E.,10, Paris, Sirey, 1974.
[545] ROMANOFF (E.): The Economic Base Model: A Very Special Case of Input-Output Analysis. J.R.S., 14,1974,121-129.
[546] SENIOR (M.L.): Approaches to Residential Location Modelling: Urban Economic Models and Some Recent Developments. Environment and Planning, A,G,1974,369-409.
[547] SENIOR (M.L.)and WILSON (A.G.): Explorations and Synthesis of Linear Programming and Spatial Interaction Models of Residential Location. Geographical Analysis, 6, 1974, 209-237.
[548] SHAKED (A.): Non-Existence of Equilibrium for the Two-Dimensional Three Firms Location Problem. R.Econ.Stud., Feb.1974, 51-56.
[549] BEAVON (K.S.O.)and MABIN (A.S.): The Losch System of Market Areas: Derivation and Extension. Geographical Analysis, 7,1975, 131-151.
[550] CADWALLADER (M.): A Behavioral Model of Consumer Spatial Decision Making. Economic Geography, 51, 1975, 339-349.
[551] CHOUKROUN (J.M.):A General Framework for the Development of Gravity-Type Distribution Models. R.S.U.E., 5,1975, 177-202.
[552] DAVIES (O.): An Examination of Properties Associated with the Isolated Producer in Two-Dimensional Spatial Markets. J.R.S., 15,1975,47-52.
[553] DAVIS (J.C.) and McCULLAGH (M.J.)eds: Display and Analysis of Spatial Data. London, John Wiley and Sons, 1975.
[554] DOMENCICH (T.A.)and McFADDEN (D.) Urban Travel Demand: a Behavioral Analysis. Amsterdam,North-Holland Publishing Cy,1975
[555] EATON (B.C.) and LIPSEY (R.G.): The Principle of Minimum Differentiation Reconsidered: Some New Developments in the Theory of Spatial Competition. R.Econ.Stud., 42,1975,27-49.
[556] FUSTIER (B.): Attraction commerciale des points de vente. Revue de l'Economie du Centre-Est, 68,1975,243-256.
[557] GOULD (P.): Les mathematiques en geographie: revolution theorique ou apparition d'un nouvel outil. Revue Internationale des Sciences Sociales, 27, 1975,319-347.
[558] GOULD (P.): Acquiring Spatial Information. Economic Geography, 51,1975,87-89.
[559] GREENHUT (J.)and GREENHUT (M.L.):Spatial Price Discrimination,Competition, and Locational Effects. Economica, 42,1975, 401-419.
[560] GREENHUT (M.L.), HWANG (M.J.), and OHTA (H.): Observations of the Shape and Relevance of the Spatial Demand Function. Econometrica, 43, 1975, 669-682.
[561] GREENHUT (M.L.)and OHTA (H.): Discriminatory and Nondiscriminatory Spatial Prices and Outputs Under Varying Market Conditions. Weltwirts. Archiv., 111, 1975,310-332.
[562] HAITES (E.F.): The Loschian Numbers as a Problem in Number Theory. Geographical Analysis, 7,1975,421-426.
[563] HOLAHAN (W.L.): The Welfare Effects of Spatial Price Discrimination. Amer.Econ.R., 65,1975,498-503.

[564] HURTER (A.P.), SCHAEFFER (M.K.), & WENDELL (R.E.): Solutions of Constrained Location Problems. Management Science, 22, 1975,51-56.
[565] ISARD (W.): A Simple Rationale for Gravity Model Type Behavior. Papers R.S.A., 35, 1975,25-30.
[566] KANEMOTO (Y.): Congestion and Cost-Benefit Analysis in Cities. Journal of Urban Economics, 2, 1975,246-264.
[567] MacKINNON (J.G.): An Algorithm for the Generalized Transportation Problem. R.S.U.E., 5,1975,445-464.
[568] MOUGEOT (M.): Theorie et Politique economiques regionales. Paris, Economica, 1975.
[569] NIJKAMP (P.): Reflexions on Gravity and Entropy Models. R.S.U.E., 5,1975, 203-225.
[570] PAELINCK (J.H.) & NIJKAMP (P.): Operational Theory and Method in Regional Economics. Farnborough, Saxon House, 1975.
[571] PARR (J.B.),DENIKE (K.G.),and MULLIGAN (G.):City-Size Models and the Economic Base: A Recent Controversy. J.R.S., 15, 1975,1-8.
[572] PONSARD (C.): Contribution a une theorie des espaces economiques imprecis. Publications Econometriques, 8,1975,1-43.
[573] PREVOT (M.): Espaces topologiques et metriques en analyse economique spatiale. University of Dijon, Unpublished Doctoral Dissertation,1975.
[574] PULLEN (M.J.): Objectives, Constraints and Spatial Behaviour. in: Environment, Man and Economic Change, Essays presented to S.H.Beaver, 371-391. Ed.Phillips and Turton, London, Longman Group Limited, 1975.
[575] RICHTER (D.K.): Existence of General Equilibrium in Multi-Regional Economics with Public Goods. International Economic Review, 16, 1975,201-221.
[576] ROUGET (B.): L'analyse spatiale en economie urbaine: Essai methodologique. University of Dijon, Unpublished Doctoral Dissertation, 1975.
[577] SMITH (T.E.): A Choice Theory of Spatial Interaction. R.S.U.E., 5,1975,137-176.
[578] SMITH (T.E.): An Axiomatic Theory of Spatial Discounting Behavior. Papers R.S.A., 35,1975,31-44.
[579] TELLIER (L.N.) and SANKOFF (D.): Gravity Models and Interaction Probabilities. J.R.S., 15,1975,317-322.
[580] THISSE (J.F.) Contribution a la theorie microeconomique spatiale. University of Liege,Unpublished Doctoral Dissertation, 1975.
[581] ZOLLER (H.G.): Housing Demand Analysis and the Spatial Factor. Universite Catholique de Louvain, Working Paper, 7507, SPUR, 1975.

FROM 1976 TO 1980

[582] BATTY (M.): Urban Modelling: Algorithms, Calibrations, Predictions. Cambridge, U.K., Cambridge University Press, 1976.
[583] BEAVON (K.S.O.): Central Place Theory: A Reinterpretation. London and New-York, Longman Inc. 1976.
[584] BEAVON (K.S.O.): A Pedagogic Approach to the Loschian System of Market Areas. Tijdschrift voor Economische en Sociale Geografie, 67,1976,29-37.

[585] BECKMANN (M.): Spatial Price Policies Revisited. Bell Journal of Economics, 7,1976, 619-630.
[586] BOVENTER (E.von): Transportation, Costs, Accessibility, and Agglomeration Economies: Centers, Subcenters, and Metropolitan Structures. Papers R.S.A., 37, 1976, 167-183.
[587] CHEVAILLER (J.C.): Affectation spatiale des ressources en presence d'effets externes. University of Dijon, Doctoral Dissertation, S.R.T.ed., 1976.
[588] COOPER (L.): An Efficient Heuristic Algorithm for the Transportation-Location Problem. J.R.S., 16,1976,309-315.
[589] EATON (B.C.): Free Entry in One-Dimensional Models: Pure Profits and Multiple Equilibria. J.R.S., 16,1976,21-33.
[590] EATON (B.C.) and LIPSEY (R.): The Non-Uniqueness of Equilibrium in the Loschian Location Model. Amer.Econ.R. 66,1976,77-93.
[591] FUJITA (M.): Toward a Dynamic Theory of Urban Land-Use. Papers R.S.A., 37,1976, 133-165.
[592] HAITES (E.F.): A Note on the General Structure of a Loschian Hierarchy. J.R.S., 16, 1976,257-260.
[593] HAY (D.A.): Sequential Entry and Entry-Deterring Strategies. Oxford Economic Papers, July 1976, 240-257.
[594] HEFFLEY (D.R.): Efficient Spatial Allocation in the Quadratic Assignment Problem. Journal of Urban Economics, 3,1976, 309-322.
[595] HORDIJK (L.) and PAELINCK (J.H.P.): Spatial Econometrics: Some Contributions. in PAELINCK (J.)ed.: Modeles en Geographie et en Economie regionale. French Speaking Regional Science Association, Proceedings 1974-1975, Rotterdam,1976.
[596] HORDIJK (L.) and PAELINCK(J.H.P.): Some Principles and Results in Spatial Econometrics. Recherches Economiques de Louvain , 3,1976,175-197.
[597] JUCKER (J.) and CARLSON (R.C.): The Simple Plant Location Problem under Uncertainty. Operations Research, 24, 1976, 1045-1055
[598] KANEMOTO (Y.): Optimum, Market and Second-Best Land Use Patterns in a Von Thunen City with Congestion. R.S.U.E., 6,1976,23-32.
[599] MacKINNON (J.G.): A Technique for the Solution of Spatial Equilibrium Models. J.R.S., 16,1976, 293-307.
[600] PAPAGEORGIOU (G.J.): Urban Residential Analysis: Spatial Consumer Behavior. Environment and Planning A, 8,1976,423-442.
[601] PAPAGEORGIOU (G.J.)ed.: Mathematical Land Use Theory. Lexington, Mass, Lexington Books, 1976.
[602] ROUGET (B.): Representations topologiques de l'espace urbain. Recherches Economiques de Louvain, 42,1976,241-254.
[603] SCHWEIZER (U., VARAIYA (P.), and HARTWICK (J.)): General Equilibrium and Location Theory. Journal of Urban Economics, 3,1976, 285-303.
[604] SCOTT (A.J.): Land and Land Rent: an Interpretative Review of the French Literature. Progress in Geography, 9,1976.
[605] SCOTT (A.J.: Land Use and Commodity Production. R.S.U.E., 6,1976, 147-160.
[606] ALAO, NURUDEEN et al.: Christaller Central Place Structures: An Introductory Statment. Evanston, Department of Geography, Northwestern University, 1977.
[607] ARPIN (O.): Theorie des places centrales et structures topologiques. University of Dijon, Unpublished Dissertation,1977.
[608] CAPOZZA (D.R.) and Van ORDER (R.): Pricing under Spatial Competition and Spatial Monopoly. Econometrica, Nov.1977.
[609] DELOCHE (R.): Theorie des sous-ensembles flous et classification en analyse economique spatiale. Revue d'Economie Politique, 3, 1977, 435-460.

[610] EATON (B.C.) and LIPSEY (R.G.): The Introduction of Space into the neoclassical Model of Value Theory. in ARTIS (M.J.) and NOBAY (A.R.)eds.: Studies in Modern Economic Analysis, Oxford, Basil Blackwell, 1977.

[611] ERLENKOTTER (D.): Facility Location with Price-Sensitive Demands: Private, Public and Quasi-Public. Management Science, 24, 1977,378-386.

[612] FISH (O.): Spatial Equilibrium with Local Public Goods, Urban Land Rent, Optimal City Size and the Tiebout Hypothesis. R.S.U.E., 7,1977,197-216.

[613] GANNON (C.A.): Product Differentiation and Locational Competition in Spatial Markets. International Economic Review, 18, 1977, 293-322.

[614] GREENBERG (J.): Existence of an Equilibrium with Arbitrary Tax Schemes for Financing Local Public Goods. Journal of Economic Theory, 16,1977,137-150.

[615] HANSEN (P.) and THISSE (J.F.): Multiplant Location for Profit Maximisation. Environment and Planning, 9,1977,63-73.

[616] HURIOT (J.M.): La formation du paysage economique. Essai sur l'affectation de l'espace. Paris, Coll.I.M.E.,15,Sirey,1977.

[617] KUENNE (R.E.): Spatial Oligopoly: Price-Location Interdependence and Social Cost in a Discrete Market Space. R.S.U.E., 7,1977, 339-358.

[618] MARSHALL (J.U.): The Construction of the Loschian Landscape. Geographical Analysis, 9, 1977,1-13.

[619] MORRILL (R.L.) and SYMONS (J.): Efficiency and Equity Aspects of Optimum Location. Geographical Analysis, 9, 1977,215-225.

[620] NIJKAMP (P.): Theory and Application of Environmental Economics. Amsterdam, North-Holland, 1977.

[621] OSLEEB (J.P.) and CROMLEY (R.G.): The Location - Production - Allocation Problem with Nonlinear Production Costs. Geographical Analysis, 9,1977,142-159.

[622] PINTO (J.V.): Launhardt and Location Theory: Rediscovery of a Neglected Book. J.R.S., 17,1977,17-29.

[623] PONSARD (C.): La region en analyse spatiale. in Comptes-Rendus du Congres sur la Methodologie de l'Amenagement et du Developpement, Universite du Quebec, Trois-Rivieres, 1977,239-248.

[624] PONSARD (C.): Hierarchie des places centrales et graphes phi-flous. Environment and Planning A,9,1977,1233-1252.

[625] RICHARDSON (H.W.): The New Urban Economics and Alternatives. London, Pion, 1977.

[626] SCHEUBEL (J.): Contribution a une theorie spatiale du consommateur. University of Dijon, Unpublished Doctoral Dissertation.1977.

[627] THISSE (J.F.) and PERREUR (J.): Relations Between the Point of Maximum Profit and the Point of Minimum Total Transportation Cost: A Restatement. J.R.S., 17,1977,227-234.

[628] WESOLOWSKY (G.O.): The Weber Problem with Rectangular Distances and Randomly Distributed Destinations. J.R.S., 17,1977,53-60.

[629] ZOLLER (H.G.): Energie, transports et structure urbaine. Recherches Economiques de Louvain, 43,1977, 323-344.

[630] ALONSO (W.): A Theory of Movement. in HANSEN (N.M.) ed.: Human Settlements. Cambridge, Mass., Ballinger, 1978.

[631] BEAVON (K.S.O.): A Comment on the Procedure for Determining the General Structure of a Loschian Hierarchy. J.R.S., 18,1978, 127-132.

[632] CAPOZZA (D.R.) and ORDER (R.Van): A Generalized Model of Spatial Competition. Amer.Econ.R., 68,1978,896-908.
[633] CHEVAILLER (J.C.): Elements d'econometrie spatiale. Coll.I.M.E., 18, Paris, Sirey, 1978.
[634] COOPER (L.): Bounds on the Weber Problem Solution under Conditions of Uncertainty. J.R.S., 18, 1978, 87-93.
[635] DE MESNARD (L.): La dominance regionale et son imprecision: traitement dans le type general de structure. Working Paper, I.M.E.32, 1978.
[636] ERLENKOTTER (D.): A Dual-Based Procedure for Uncapacitated Facility Location. Operations Research, 16,1978,992-1009.
[637] FUNCK (R.) and PARR (J.B.): The Analysis of Regional Structure: Essays in Honour of August Losch. London, Pion Limited,1978.
[638] FUSTIER (B.): L'attraction des points de vente dans des espaces precis et imprecis. Working Paper, IME, 30, 1978.
[639] GREENBERG (J.): Pure and Local Public Goods: a Game Theoretic Approach. in: Essays in Public Economics, ed.by Agnar Sandmo, Lexington, Massachusetts, Heath and Co, 1978,49-78.
[640] GREENHUT (M.L.): Impact of Distance on Microeconomic Theory. Manchester School of Economic and Social Studies. 46,1978, 17-40.
[641] HAINING (R.P.) : Interaction Modelling on Central Place Lattices. J.R.S., 18, 1978, 217-228.
[642] HOLAHAN (W.L.): Spatial Competition Versus Spatial Monopoly. Journal of Economic Theory, 18, 1978,156-170.
[643] HORN (R.J.) and PRESCOTT (J.R.): Central Place Models and the Economic Base: Some Empirical Results. J.R.S., 18,1978,229-241.
[644] HURIOT (J.M.): Utilisation de l'espace et dynamique economique. Revue d'Economie Regionale et Urbaine, 2,1978,119-148.
[645] JONES (A.P.), McGUIRE (W.J.), and WITTE (A.D.): A Reexamination of Some Aspects of von Thunen's Model of Spatial Location. J.R.S., 18,1978,1-15.
[646] KENNEDY (P.) and COPES (P.): Product Differentiation and Centralization of Production, J.R.S., 18, 1978, 323-335.
[647] LEVEN (C.L.): Growth and NonGrowth in Metropolitan Areas and the Emergence of Polycentric Metropolitan Form. Papers R.S.A., 41, 1978, 101-112.
[648] MARSHALL (J.U.): On the Structure of the Loschian Landscape. J.R.S., 18, 1978, 121-125.
[649] MARTIN (R.L.), THRIFT (N.J.), and BENNETT (R.J.) eds: Towards the Dynamic Analysis of Spatial Systems. London, Pion Limited, 1978.
[650] MILLER (S.M.) and JENSEN (O.W.): Location and the Theory of Production. R.S.U.E., 8,1978, 117-128.
[651] MILLERON (J.C.): L'equilibre des agents economiques dans l'espace. Revue Economique, 29,1978,237-260.
[652] MOUGEOT (M.): The Welfare Foundations of Regional Planning: General Equilibrium and Pareto Optimality in a Spatial Economy. R.S.U.E., 8,1978,175-194.
[653] PAELINCK (J.H.P.): Spatial Econometrics. Economic Letters, 1,1978, 59-63.
[654] PAELINCK (J.H.P.): Une theorie des seuils de croissance regionaux. in Seuils d'efficacite de la planification et de l'action regionale. I.D.E.A., Mons, 1978,101-107.
[655] PAELINCK (J.H.P.) and ZOLLER (H.G.): Some Spatial Consumer Theory. Working Paper, 7805,S.P.U.R.,Universite Catholique de Louvain, 1978.
[656] PONSARD (C.): Esquisse de simulation d'une economie regionale: l'apport de la theorie des systemes flous.in Melanges economiques en hommage a Pierre Moran. Paris,Economica,1978, 24-38.

[657] RICHTER (D.K.): Existence and Computation of a Tiebout General Equilibrium. Econometrica, 46,1978,779-805.
[658] RICHTER (D.K.)): The Computation of Urban Land Use Equilibria. Journal of Economic Theory, 19,1978,1-27.
[659] SENINGER (S.F.): Expenditure Diffusion in Central Place Hierarchies: Regional Policy and Planning Aspects. J.R.S., 18, 1978, 243-261.
[660] TRANQUI (P.): Les regions economiques floues: application au cas de la France. Coll.I.M.E.,16, Paris, Sirey,1978
[661] TULKENS (H.): Dynamic Processes for Public Goods: An Institution-Oriented Survey. Journal of Public Economics, 9,1978, 163-201.
[662] WOODERS (M.): Equilibria, the Core and Jurisdiction Structures in Economies with a Local Public Good. Journal of Economic Theory, 18, 1978, 328-348.
[663] ANCOT (J.P.): Une approche par analyse discriminante a des problemes de seuils regionaux et d'analyse de localisation: application a des donnees espagnoles. Recherches Economiques de Louvain, 45, 1979, 281-297.
[664] ANCOT (J.P.) and PAELINCK (J.H.P.): A Discriminant Analysis Approach to Regional Threshold Problems with an Application to Spanish Data. Papers R.S.A., 42,1979,139-152.
[665] BEAVON (K.S.O.): The Losch Constraints - Again. J.R.S., 19,1979, 505-509.
[666] BEGUIN (H.): City Size and Urban Hierarchy Models. Geographical Analysis, Jan.1979.
[667] BEGUIN (H.): Urban Hierarchy and the Rank-Size Distribution. Geographical Analysis, 11,1979,149-164.
[668] BEGUIN (H.): Methodes d'analyse geographique quantitative. Paris, Librairies Techniques, 1979.
[669] BEGUIN (H.) and THISSE (J.F.): An Axiomatic Approach to Geographical Space. Geographical Analysis, 11, 1979,325-341.
[670] BLAUG (M.): The German Hegemony of Location Theory: a Puzzle in the History of Economic Thought. History of Political Economy, 11,1979, 21-29.
[671] BONIVER (V.): Un apercu de la nouvelle economie urbaine. Revue d'Economie Regionale et Urbaine. 3-4, 1979,327-362.
[672] DANDRI (G.): Dove e Perche. Territorio e Costruzioni nelle Dottrine Economiche. Napoli, Guida Editori, 1979.
[673] D'ASPREMONT (C.), JASKOLD-GABSZEWICZ (J.) and THISSE (J.F.): On Hotelling's "Stability in Competition". Econometrica, 47,1979,1145-1150.
[674] EATON (B.C.) and LIPSEY (R.G.): Comparison Shopping and the Clustering of Homogeneous Firms. J.R.S., 19,1979,421-435.
[675] ELLICKSON (B.): Competitive Equilibrium with Local Public Goods. Journal of Economic Theory, 21, 1979,46-61.
[676] FUSTIER (B.): Les interactions spatiales en economie. Coll.I.M.E., 21, Paris, Sirey, 1979.
[677] GUESNERIE (R.) and ODDOU (C.): On Economic Games which Are Not Necessarily Superadditive: Solution Concepts and Application to a Local Public Good Problem with Few Agents. Economic Letters, 3,1979, 301-305.
[678] HANSEN (P.), PEETERS (D.) and THISSE (J.F.): An Efficient Algorithm for the Constrained Weber Problem. Working Papers, SPUR, 7907, June 1979.

[679] HORDIJK (L.) and PAELINCK (J.H.P.): Spatial Econometrics: Some Further Results. 2nd International Colloquium of Applied Econometrics, Modeles regionaux et modeles regionaux-nationaux, Nice, 1975, Proceedings, Paris, Cujas, 1979, 275-285.

[680] ISARD (W.) and LIOSSATOS (P.): Spatial Dynamics and Optimal Space-Time Development. New-York, Amsterdam, Oxford, North-Holland Publishing Company, 1979.

[681] KLAASSEN (L.H.), PAELINCK (J.H.P.) and VERSTER (A.C)): Deux problemes dans la modelisation du comportement de deplacement. Les Cahiers Scientifiques de la Revue Transport, 1979,72-93.

[682] KLAASSEN (L.H.), PAELINCK (J.H.P.) and WAGENAAR (S.): Spatial Systems A General Introduction. Farnborough, Saxon House,1979.

[683] LEA (A.C.): Welfare Theory, Public Goods and Public Facility Location. Geographical Analysis, 11,1979,217-239.

[684] MARSHALL (J.U.): Losch Revisited - Again. J.R.S., 19,1979,501-503.

[685] MATHUR (V.K.): Some Unresolved Issues in the Location Theory of the Firm. Journal of Urban Economics, 6,1979,299-318.

[686] MULLIGAN (G.F.): Additional Properties of a Hierarchical City-Size Model. J.R.S., 19,1979,57-66.

[687] OKABE (A.): An Expected Rank-Size Rule: A Theoretical Relationship Between the Rank-Size Rule and City Size Distributions. R.S.U.E., 9,1979,21-40.

[688] PAELINCK (J.H.P.) and KLAASSEN (L.H.): Spatial Econometrics. Farnborough, Saxon House, 1979.

[689] PAELINCK (J.H.P.) and TACK (D.): Distance Interaction : Rationale, Estimation, and Computing. Netherlands Economic Institute , Foundations of Empirical Economic Research, Rotterdam, 4,1979.

[690] PONSARD (C.): Economie urbaine et espaces metriques. Sistemi Urbani, 1,1979,123-135.

[691] PONSARD (C.): On the Imprecision of Consumer's Spatial Preferences. Papers R.S.A., 42,1979,59-71.

[692] ROUGET (B.): Espaces topologiques et espaces urbains. Sistemi Urbani, 2,1979,39-47.

[693] SCOTT (A.J.) : Commodity Production and the Dynamics of Land-Use Differentiation. Urban Studies, 16, 1979,95-104.

[694] BEAUMONT (J.R.): Spatial Interaction Models and the Location-Allocation Problem. J.R.S., 20, 1980, 37-50.

[695] FUSTIER (B.): Contribution a l'analyse spatiale de l'attraction imprecise. Sistemi Urbani, 2-3, 1980, 271-282.

[696] GRAITSON (D.): On Hotelling's "Stability in Competition" Again. Economic Letters, 6,1980,1-6.

[697] HAINING (R.): Intraregional Estimation of Central Place Population Parameters. J.R.S., 20, 1980, 365-375.

[698] HANJOUL (P.): Facility Location under Particular Allocation Rules. Workshop on Location and Distribution Management, Bruxelles, 1980.

[699] HANSEN (P.), PEETERS (D.) and THISSE (J.F.): Location of Public Services: a Selective Method-Oriented Survey. Annals of Public and Cooperative Economy, 51,1980, 9-51.

[700] HANSEN (P.), PERREUR (J.) and THISSE (J.F.): Location Theory, Dominance and Convexity: Some Further Results. Operations Research, 28,1980, 1241-1250.

[701] JURION (B.): Contributions a l'economie publique locale. Quelques applications aux fusions de communes. University of Liege, Unpublished Doctoral Dissertation., 1980.

[702] MULLIGAN (G.F.): The Effects of Multiplier Shifts in a Hierarchical City-Size Model. R.S.U.E., 10, 1980, 77-90.
[703] OHTA (H.): Spatial Competition, Concentration and Welfare. R.S.U.E., 10, 1980, 3-16.
[704] PEETERS (D.) and ZOLLER (H.G.): Quelques applications de la theorie des graphes en geographie. in HANSEN (P.) and DE WERRA (D.) eds: Regards sur la theorie des graphes. Lausanne, Presses Polytechniques Romandes, 1980.
[705] PESTIEAU (P.): Fiscal Mobility and Local Public Goods. A Survey of the Empirical and Theoretical Studies of the Tiebout Model. Research Programm Papers, S.P.U.R., 6,1980.
[706] PONSARD (C.): Fuzzy Economic Spaces. First World Regional Science Congress, Cambridge, Mass.,June 1980, Working Paper , I.M.E.,43,1980 (Russian trans.,forthcoming).
[707] SHEPPARD (E.S.): Location and the Demand for Travel. Geographical Analysis, 12,1980,111-128.
[708] WOODERS (M.): The Tiebout Hypothesis: Near Optimality in Local Public Good Economies. Econometrica, 48,1980,1467-1485.

AFTER 1980

[709] ANCOT (J.P.), KUIPER (J.H.), and PAELINCK (J.H.P.): Econometrie spatiale: une synthese decennale. Rotterdam, Netherlands Economic Institute and University Erasmus. August 1981.
[710] ANCOT (J.P.) and PAELINCK (J.H.P.): Recent Research in Spatial Econometrics. in GRIFFITH (D.) and MACKINNON (R.)eds: Dynamic Spatial Models. Sijthoff and Noordhoff, Alphen a/d Rijn and Rochville, 1981, 344-364.
[711] BEGUIN (H.) and PEETERS (D.): Urbanization in Some Hierarchical Urban Models. R.S.U.E., 11,1981.
[712] BERGLAS (E.) and PINES (D.): Clubs, Local Public Goods and Transportation Models: A Synthesis. Journal of Public Economics , 15,1981, 141-162.
[713] BEWLEY (T.): A Critique of Tiebout's Theory of Local Public Expenditures. Econometrica, 49,1981,713-740.
[714] CARRUTHERS (N.): Central Place Theory and the Problem of Aggregating Individual Location Choices. J.R.S., 21,1981,243-261.
[715] D'ASPREMONT (C.), JASKOLD-GABSZEWICZ (J.) and THISSE (J.F.): On Hotelling's "Stability in Competition" (continued). Working Papers, SPUR, 8102, June 1981.
[716] ESWARAN (M.), KANEMOTO (Y.) and RYAN (D.): A Dual Approach to the Locational Decision of the Firm. J.R.S., 21,1981,469-490.
[717] GRONBERG (T.) and MEYER (J.): Transport Inefficiency and the Choice of Spatial Pricing Mode. J.R.S., 21,1981,541-549.
[718] HANSEN (P.), PEETERS (D.), RICHARD (D.) and THISSE (J.F.): The Rawls Locational Problem Revisited. Research Program Papers , S.P.U.R., June 1981.
[719] HANSEN (P.), PEETERS (D.) and THISSE (J.F.): Some Localization Theorems for a Constrained Weber Problem. J.R.S. 21, 1981,103-115.
[720] HANSEN (P.), PEETERS (D.) and THISSE (J.F.): Constrained Location and the Weber-Rawls Problem. Annals of Discrete Mathematics, 11,1981,147-166.
[721] HANSEN (P.) and THISSE (J.F.): The Generalized Weber-Rawls Problem. Research Program Papers, S.P.U.R., Feb.1981.

[722] HANSEN (P.) and THISSE (J.F.): Outcomes of Voting and Planning: Condorcet, Weber and Rawls Locations. Journal of Public Economics, 16,1981, 1-15.
[723] HANSEN (P.) and THISSE (J.F.): The Weber Problem Revisited. Research Program Papers, S.P.U.R., 10,1981.
[724] HANSEN (P.) and THISSE (J.F.): Recent Advances in Continuous Location Theory. Sistemi Urbani, 1982,(forthcoming).
[725] HANSEN (P.), THISSE (J.F.) and HANJOUL (P.): Simple Plant Location under Uniform Delivered Pricing. European Journal of Operational Research, 6, 1981, 94-103.
[726] HANSEN (P.), THISSE (J.F.) and WENDELL (R.E.): Weber, Condorcet and Plurality Solutions to Network Location Problems. Research Program Papers, S.P.U.R., 14,1981.
[727] HEBERT (R.F.): Richard Cantillon's Early Contributions to Spatial Economics. Economica, 48,1981,71-77.
[728] HOLAHAN (W.L.) and SCHULER (R.E.): Competitive Entry in a Spatial Economy: Market Equilibrium and Welfare Implications. J.R.S., 21,1981, 341-357.
[729] HURIOT (J.M.): Rente fonciere et modele de production. Environment and Planning, A,13,1981, 1125-1149.
[730] JUEL (H.): Bounds in the Location-Allocation Problem. J.R.S., 21, 1981, 277-282.
[731] MIRON (J.R.) and SKARKE (P.): Non-Price Information and Price Sustainability in the Koopmans-Beckmann Problem. J.R.S., 21, 1981, 117-122.
[732] MULLIGAN (G.F.): The Urbanization Ratio and the Rank-Size Distribution: A Comment. J.R.S., 21,1981, 283-285.
[733] PONSARD (C.): An Application of Fuzzy Subsets Theory to the Analysis of the Consumer's Spatial Preferences. Fuzzy Sets and Systems, 5,1981, 235-244.
[734] PONSARD (C.): L'equilibre spatial du consommateur dans un contexte imprecis. Sistemi Urbani, 1-2, 1981,107-133.
[735] PONSARD (C.): Comportements flous et equilibres spatiaux partiels. Revue d'Economie Politique, 6, 1981,920-930.
[736] ROUGET (B.): Equilibre spatial du consommateur. Une analyse multidimensionnelle. University of Dijon, Unpublished Doctoral Dissertation, 1981.
[737] ROUGET (B.): Images de la ville et topologie floue. Sistemi Urbani, 1-2, 1981,93-105.
[738] SHILONY (Y.): Hotelling's Competition with General Customer Distribution. Economic Letters, 1981.
[739] SMITH (M.J.de): Optimum Location Theory- Generalizations of Some Network Problems and Some Heuristic Solutions. J.R.S. 21, 1981, 491-505.
[740] THISSE (J.F.) and PAPAGEORGIOU (G.J.): Reconciliation of Transportation Costs and Amenities as Location Factors in the Theory of the Firm. Geographical Analysis, 13, 1981,189-195.
[741] WENDELL (R.E.) and McKELVEY (R.D.): New Perspectives in Competitive Location Theory. European Journal of Operational Research, 6,1981,174-182.
[742] WHITMORE (Jr.H.W.): Plant Location and the Demand for Investment: A Theoretical Analysis. J.R.S., 21,1981, 89-101.
[743] DELOCHE (R.): Economie Publique Locale et Relations Institutionnelles: Une Approche en Termes d'Efficacite, de Justice et de Stabilite. University of Besancon, Doctoral Dissertation, Working Paper, G.R.E.M., 8203,1982.

[744] GRAITSON (D.): Spatial Competition a la Hotelling: a Selective Survey. Journal of Industrial Economics, 1982 (forthcoming).
[745] GUIGOU (J.L.): La rente fonciere. Les theories et leur evolution depuis 1650. Paris, Economica, 1982.
[746] LOUVEAUX (F.), THISSE (J.F.) and BEGUIN (H.): Location Theory and Transportation Costs. R.S.U.E., 12, 1982.
[747] PAELINCK (J.H.P.) and ZOLLER (H.G.): A Discrete Model of Residential Choice. in PAELINCK (J.H.P.), ed.: Formal Spatial Economic Analysis. Farnborough, U.K., Gower Press, 1982.
[748] PEETERS (D.): Methodes de localisation des services publics. 1982. (forthcoming).
[749] PONSARD (C.): Producer's Spatial Equilibrium with a Fuzzy Constraint. European Journal of Operational Research, 10,1982,302-313.
[750] PONSARD (C.): Partial Spatial Equilibria with Fuzzy Constraints. J.R.S., 22, 1982, 159-175.
[751] PONSARD (C.) and TRANQUI (P.): La regionalisation de l'economie europeenne. Congres International des Economistes de Langue Francaise, Florence, 1982, Working Paper, IME, 57, 1982 and Revue d'Economie Politique (forthcoming)
[752] THISSE (J.F.) and ZOLLER (H.G.): Some Notes on Public Facility Location. in THISSE (J.F.) & ZOLLER (H.G.)eds: Locational Analysis of Public Facilities. Amsterdam, North-Holland American Elsevier, (forthcoming).

Author Index

A number refers to a single page; the underlined numbers refer to a chapter, a section or a subsection.

ACKLEY (G.): 90
ALAO : 109
ALLAIS (M.): 118
ALLEN (W.B.): 115
ALONSO (W.) : 101 - 103 - 105 - 122 - 125
ANAS (A.) : 102
ARPIN (O.) : 122
AUERBACH (F.) : 110

BALINSKI (M.L.) : 106
BEAUMONT (J.R.) : 105
BEAVON (K.S.O.) : 109
BECHDOLT (B.V.) : 115
BECKMANN (M.J.) : 11 - 12 - 94 - 95 - 96 - 102 - 108 - 110 - 111 - 115 - 116 - 117 - 121
BEGUIN (H.) : 110 - 125
BERRY (B.J.L.) : 109
BLACK (J.D.) : 21
BODENHORN (D.) : 103
BORTKIEWICZ (L.) : 33 - 45
BOUDEVILLE (J.R.) : 99
BOUNON (C.) : 121
BOVENTER (E.von) : 116
BULOW (F.) : 13 - 97
BURTON (R.) : 118

CADWALLADER (M.) : 115
CANTILLON (R.) : 6 - 7 - 8 - 9 - 10 - 12 - 99
CARROTHERS (G.A.P.) : 113
CARRUTHERS (N.) : 112
CHAMBERLIN (E.H.) : 12 - 36 - 41 - 43 - 45 - 61 - 73 - 76 - 91 - 107
CHEVAILLER (J.C.) : 117
CHRISTALLER (W.) : 36 - 39 - 40 - 41 - 74 - 95 - 100 - <u>108 - 112</u> - 117
CHRISTOFIDES (N.) : 105
CLARK (C.) : 100 - 110
CLARK (J.B.) : 18 - 42
CONDILLAC (A.de) : 8 - 10
CONVERSE (P.D.) : 111
COOPER (L.) : 105 - 106 - 126
COPES (P.) : 108
CROMLEY (R.G.) : 105
CURRY (L.) : 108

DACEY (M.F.) : 109 - 110 - 111
DANDRI (G.) : 5
DANTZIG (G.B.) : 93
D'ASPREMONT (C.) : 108
DEAN (W.H.) : 61 - 96
DELOCHE (R.) : 119 - 124
DE MESNARD (L.) : 124
DENIKE (K.G.) : 110
DENNISON (S.R.) : 62 - 63 - 64
DEVLETOGLOU (N.E.) : 107
DEWEY (R.L.) : 93
DICKINSON (H.O.) : 100
DOCKES (P.) : 5
DUNN (E.S.) : 114
DZIEWONSKI (K.) : 110

EATON (B.C.) : 108
EILON (S.) : 105
EMERSON (D.L.) : 103
ENGLANDER (O.) : 21 - 36 - 37 - 39 - 45
ENKE (S.) : 91 - 96 - 116
EPSTEIN (R.C.) : 45
ERLENKOTTER (D.) : 106
ESWARAN (M.) : 103

FANO (P.L.) : 109
FETTER (F.A.) : 21 - 41 - 42 - 49 - 51 - 60 - 89 - 90 - 97 - 150
FLORES (E.) : 4
FREUTEL (G.) : 94 - 114
FURLAN (V.) : 33
FUSTIER (B.) : 115 - 121 - 124

GADREAU (M.) : 121
GAMBINI (R.) : 121
GANNON (C.A.) : 107 - 112
GARRISON (W.L.) : 109
GIERSCH (H.) : 96
GOLDMAN (T.A.) : 107 - 116
GOLDNER (W.) : 92
GOLOB (T.F.) : 115
GOULD (P.) : 112
GREENHUT (M.L.) : 97 - 117
GRONBERG (T.) : 112
GROTEWOLD (A.) : 100
GUIGOU (J.L.) : 121 - 122
GUSTAFSON (R.L.) : 115

HAINING (R.P.) : 110
HAITES (E.F.) : 109
HAKIMI (S.L.) : 103
HALL (F.S.) : 19 - 49
HANSEN (P.) : 104 - 105 - 106
HAPRIS (B.) : 115
HARTWICK (J.M.) : 107 - 114
HARTWICK (P.) : 107
HAWTREY (R.G.) : 37
HEBERT (R.F.) : 12
HERBERT (J.D.) : 101
HITCHCOCK (F.) : 93 - 96 - 105
HOLAHAN (W.L.) : 111
HOOVER (E.M.) : 61 - 64 - 76 - 91 - 96 - 97
HORN (R.J.) : 110
HOTELLING (H.) : 43 - 49 - 51 - 52 - 57 - 59 - 60 - 61 - 67 - 71 -
    89 - 90 - 97 - 100 - 105 - <u>106 - 108</u> - 111 - 127 -
    153 - 157 - 174
HUFF (D.L.) : 114 - 121
HURIOT (J.M.) : 100 - 101 - 104 - 121 - 122
HURTER (A.P.) : 104
HYSON (C.D.) : 90
HYSON (W.P.) : 90

ISARD (W.) : 3 - 12 - 35 - 72 - 92 - 93 - 94 - 95 - 99 - 103 - 105 -
    107 - 111 - 114 - 116 - 117

JACOT (S.P.) : 107
JASKOLD - GABSZEWICZ (J.) : 108
JENKS (G.F.) : 121
JENSEN (O.W.) : 103
JONASSON (O.) : 42 - 45 - 54
JONES (A.P.) : 44 - 100
JUDGE (G.G.) : 117
JUEL (H.) : 106

KANEMOTO (Y.) : 101 - 103
KATZ (I.N.) : 105
KATZMAN (M.T.) : 100
KEIR (M.) : 21
KENNEDY (P.) : 108
KHALILI (A.) : 103
KLAASSEN (L.H.) : 119
KOHL (J.G.) : 39
KOOPMANS (T.C.) : 93 - 96 - 116
KUENNE (R.E.) : 92 - 106 - 108

LAUNHARDT (W.) : 20 - 21 - 25 - 42 - 49 - 51 - 52 - 54 - 57 - 59 -
                 60 - 68 - 77 - 89 - 90 - 136 - 153
LAV (M.R.) : 111
LEFEBER (L.) : 107 - 116
LEONTIEF (W.W.) : 3 - 94 - 95
LERNER (A.P.) : 61 - 90 - 107
LEWIS (W.A.) : 92
LIPSEY (R.G.) : 108
LONG (W.H.) : 112 - 123
LOSCH (A.) : 3 - 6 - 10 - 12 - 23 - 31 - 36 - 41 - 45 - 48 - 60 - 61 -
             64 - 65 - 88 - 89 - 91 - 92 - 93 - 94 - 95 - 96 - 99 -
             100 - 108 - 112 - 116 - 117 - 120 - 122 - 127 -
             175 - 195
LOTKA (A.J.) : 110
LOVELL (M.) : 108

MANNE (A.S.) : 106
MARSCHAK (T.) : 96
MARSHALL (A.) : 12 - 18 - 19 - 21 - 34 - 44 - 64 - 91
MARSHALL (J.U.) : 109
MATHUR (V.K.) : 103 - 115
McGUIRE (W.J.) : 94 - 100
McKINNON (J.G.) : 106
McLAUGHLIN (G.E.) : 62 - 64
McPHERSON (J.C.) : 110
MEDVEDKOV (Y.V.) : 109
MEYER (F.) : 65
MEYER (J.) : 112
MEYER - LINDEMANN (H.U.) : 97
MIKSCH (L.) : 72 - 91 - 93 - 95
MILLER (S.M.) : 103
MILLS (E.S.) : 111
MOORE (H.L.) : 17 - 134
MORAN (P.) : 99
MOSES (L.N.) : 103 - 105
MOUGEOT (M.) : 116 - 117
MULLIGAN (G.F.) : 110
MUTH (R.F.) : 101

NEFF (P.) : 94
NIEDERCORN (J.H.) : 115
NIJKAMP (P.) : 103
NURUDEEN : 109

OHLIN (B.) : 3 - 12 - 41 - 44 - 45 - 49 - 72 - 80 - 87 - 95
OKABE (A.) : 110
OORT (C.J.) : 118
OSLEEB (J.P.) : 105

PAELINCK (J.H.P.) : 102 - 103 - 119
PALANDER (T.) : 3 - 20 - 21 - 33 - 36 - 43 - 45 - 47 - 60 - 61 - 65 -
           66 - 68 - 77 - 78 - 87 - 90 - 93 - 96 - 148 - 175
PAPAGEORGIOU (G.J.) : 123
PARR (J.B.) : 110
PEETERS (D.) : 104 - 106
PERREUR (J.) : 103 - 104 - 121 - 122
PICK (G.) : 105
PONSARD (C.) : 12 - 99 - 121 - 122 - 123 - 124 - 148
POPESCU (O.) : 12 - 97
PRED (A.) : 109
PREDOHL (A.):3 - 31 - 33 - 36 - 37 - 49 - 65 - 68 - 87 - 92 - 97 - 104
PRESCOTT (J.R.) : 110

REILLY (W.J.) : 114
REITER (S.) : 93
REYNOLDS (L.G.) : 92
RICHARD (D.) : 104
RITSCHL (H.) : 36 - 37 - 90 - 97
ROBINSON (A.) : 44 - 64 - 90
ROBINSON (E.A.G.) : 44
ROGERS (J.D.) : 105
ROMANOFF (E.) : 110
ROSCHER (W.) : 14 - 20 - 49 - 97
ROSS (E.A.) : 19 - 49
ROUGET (B.) : 122 - 123 - 124
RUSHTON (G.) : 112
RYAN (D.) : 103
RYDELL (C.P.) : 112

SAKASHITA (N.) : 103
SAMUELSON (P.A.) : 96 - 116 - 118
SARLY (R.M.) : 112
SCHAFFLE (A.) : 20 - 39 - 49 - 97
SCHARLIG (A.) : 121
SCHEUBEL (J.) : 123
SCHILLING (A.) : 42 - 49 - 51 - 54
SCHMIDT (P.H.) : 39
SCHNEIDER (E.) : 49 - 90 - 149
SCHNEIDER (M.) : 115
SCHOOLER (E.W.) : 107
SCHRAMM (G.) : 115
SCHULER (R.E.) : 111
SCOTT (A.J.) : 99 - 105 - 108
SENIOR (M.L.) : 100
SERCK - HANSSEN (J.) : 117
SHAKED (A.) : 108
SINGER (H.W.) : 61 - 107 - 110
SIRKIN (G.) : 110

SMITH (A.) : 3 - 11 - 12 - 14 - 49
SMITH (M.J.de) : 105
SMITH (T.E.) : 105
SMITHIES (A.) : 61 - 90 - 91 - 97 - 106 - 107 - 108
SMOLENSKY (E.) : 118
SOLAND (R.M.) : 106 - 108
SOMBART (W.) : 97 - 110
SPENGLER (J.J.) : 12
SRAFFA (P.) : 43 - 101
STACKELBERG (H.von) : 78 - 93
STEUART (J.) : 9 - 10
STEVENS (B.H.) : 101 - 107 - 112
STEWART (C.) : 110
STOLLSTEIMER (J.F.) : 106
STOUFFER (S.A.) : 114
STRODTBECK (F.) : 114

TACK (D.) : 102
TAKAYAMA (T.) : 117
TEITZ (M.B.) : 107 - 118
TELLIER (L.N.) : 105
TELSER (L.G.) : 108
TERMOTE (M.) : 125
THISSE (J.F.) : 103 - 104 - 105 - 106 - 108 - 118 - 122
THUNEN (J.H.von) : 3 - 5 - 7 - 8 - 10 - 11 - 12 - 13 - 18 -
                   19 - 20 - 21 - 23 - 24 - 30 - 31 - 33 - 35 - 39 -
                   45 - 49 - 50 - 54 - 67 - 68 - 69 - 70 - 100 - 102 -
                   106 - 117 - 120 - 122 - 123 - 127 - 131 - 136 -
                   177 - 179
TIDEMAN (N.) : 118
TIEBOUT (C.M.) : 110 - 118
TINBERGEN (J.) : 109
TOBLER (W.R.) : 112
TRANQUI (P.) : 124
TRIAS - FARGAS (R.) : 4

ULLMAN (E.L.) : 41 - 110
URE : 19

VERGIN (R.C.) : 105
VICKERY (W.S.) : 107
VINING (R.) : 94 - 95

WARNTZ (W.) : 114
WATSON - GANDY (C.D.T.) : 105
WEBBER (M.J.) : 122
WEBER (A.) : 3 - 10 - 19 - 20 - 21 - 23 - 31 - 33 - 34 - 35 - 36 -
             37 - 39 - 42 - 44 - 49 - 50 - 54 - 55 - 62 - 63 -
             64 - 68 - 69 - 72 - 80 - 96 - 100 - 102 - 106 -
             108 - 122 - 125 - 127 - 136 - 147
WEIGMANN (H.) : 37 - 38 - 39
WENDELL (R.E.) : 104
WESOLOWSKY (G.O.) : 105
WHITMORE (Jr.H.W.) : 105

WILSON (A.G.) : 1
WINGO (L.J.) : 101
WINSTEN (C.B.) : 94
WITTE (A.D.) : 100
WITZGALL (C.) : 104
WOODWARD (R.S.) : 103

ZIPF (G.K.) : 110 - 113
ZOLLER (H.G.) : 102 - 118 - 123

Economics from Springer

Y. Murata
Optimal Control Methods for Linear Discrete-Time Economic Systems

1982. 2 figures. X, 202 pages.
ISBN 3-540-90709-2

Here is a comprehensive, self-contained volume on methods of stabilizing linear dynamical systems in discrete-time variables, covering certainty and uncertainty cases in various informational systems.

B. Felderer
Wirtschaftliche Entwicklung bei schrumpfender Bevölkerung

Eine empirische Untersuchung

1983. 31 Abbildungen. X, 306 Seiten.
ISBN 3-540-12408-X

Die vorliegende Arbeit untersucht die volkswirtschaftlichen Konsequenzen schrumpfender Bevölkerungen. Dazu werden Simultansmodelle verwendet, deren Ergebnisse allgemeine Schlußfolgerungen über die langfristigen Auswirkungen schrumpfender Bevölkerungen auf die wichtigsten makroökonomischen Variablen erlauben. Insbesondere wird die Abhängigkeit von Pro-Kopf-Einkommen, Pro-Stunden-Einkommen, Investitionen, Kapitalstock, Staatsausgaben, Steuerquote, Staatsverschuldung, Rentenversicherung und gesetzliche Krankenversicherung von der Bevölkerungsentwicklung erörtert.

H. Dyckhoff
Handelsgewinne rohstoffarmer Industrieländer und rohstoffreicher Entwicklungsländer

Eine spieltheoretische Analyse

1983. 27 Abbildungen. XII, 256 Seiten.
ISBN 3-540-12122-6

In diesem Buch wird der Versuch unternommen, den vertikalen Welthandel zwischen Industrie- und Rohstoffländern theoretisch zu analysieren. Kernfrage ist dabei die Frage nach der Aufteilung der Handelsgewinne. Mit Hilfe von Lösungskonzepten der Spieltheorie gelingt es, ein theoretisches Konzept für eine „faire" Aufteilung zu schaffen, mit dem die „Ausbeutung" der Handelspartner vermieden werden kann.

H.-B. Schäfer
Landwirtschaftliche Akkumulationslasten und industrielle Entwicklung

Analyse und Beschreibung entwicklungspolitischer Optionen in dualistischen Wirtschaften

1983. XVI, 345 Seiten.
ISBN 3-540-12234-6

Das Werk behandelt die verschiedenen Theorien über die Rolle der Landwirtschaft in den Anfangsstadien wirtschaftlicher Entwicklung. Es bietet damit eine umfassende Analyse der Vorgänge im Agrarsektor, die Einflüsse auf das industrielle Wachstum nehmen können. Es stellt die Grundmuster der wirtschaftspolitischen Optionen dar, die sich auf die Gestaltung des Verhältnisses zwischen Landwirtschaft und Industrie beziehen, und beschreibt Entwicklungsphasen in fünf Ländern, während derer sich die Landwirtschaft als strategischer Sektor erwies.

Springer-Verlag
Berlin
Heidelberg
New York
Tokyo

Lecture Notes in Economics and Mathematical Systems

Managing Editors: M. Beckmann, W. Krelle

This series reports new developments in (mathematical) economics, econometrics, operations research, and mathematical systems, research and teaching – quickly, informally and at a high level.

Volume 216
H. H. Müller, Zürich, Switzerland
Fiscal Policies in a General Equilibrium Model with Persistent Unemployment
1983. VI, 92 pages.
ISBN 3-540-12316-4

Contents: Introduction. - The General Model: Formulation of the General Model. Equilibrium. Existence of Equilibria. The Set of Equilibria and Relative Pareto Optima. First Best Pareto Optima. - The Three Commodity Model with Growth of Population: Formulation of the Model. Equilibrium. Existence of Equilibria. Further Results. Comparative Statics. Examples. - Appendix (Proofs). - References.

Volume 214
M. Faber, H. Niemes, G. Stephan, Universität Heidelberg, Germany
Entropie, Umweltschutz und Rohstoffverbrauch
Eine naturwissenschaftlich ökonomische Untersuchung

Unter Mitarbeit von L. Freytag
1983. IX, 181 Seiten.
ISBN 3-540-12297-4

Ziel dieses Buches ist es, Umweltschutz und Rohstoffverbrauch in einem umfassenderen Zusammenhang, als dies bisher geschehen ist, zu untersuchen. Dabei werden nicht nur ökonomische, sondern auch naturwissenschaftliche Aspekte berücksichtigt. Mit ihrer interdisziplinären Vorgehensweise versuchen die Autoren, die Irreversibilität ökonomischer Prozesse und damit deren zeitlichen Verlauf stärker als bisher zu berücksichtigen. Umweltschutz und Rohstoffverbrauch werden erst separat, dann gemeinsam untersucht.

Springer-Verlag
Berlin Heidelberg
New York Tokyo

Volume 213
Aspiration Levels in Bargaining and Economic Decision Making
Proceedings of the Third Conference on Experimental Economics, Winzenhohl, Germany, August 29 – September 3, 1982
Editor: R. Tietz, University of Frankfurt am Main, Germany
1983. VIII, 406 pages.
ISBN 3-540-12277-X

The experimental results and the new approaches presented herein aid in the understanding of human decision behavior and suggest how to react accordingly.

Volume 212
R. Sato, Brown University, Providence, RI, USA;
T. Nôno, Fukuoka University, Munakata, Fukuoka, Japan
Invariance Principles and the Structure of Technology
1983. V, 94 pages.
ISBN 3-540-12008-4

Contents: Introduction. - Lie Group Methods and the Theory of Estimating Total Productivity. - Invariance Principle and "G-Neutral" Types of Technical Change. - Analysis of Production Functions by "G-Neutral" Types of Technical Change. - Neutrality of Inventions and the Structure of Production Functions. - References.

Volume 210
Technology, Organization and Economic Structure
Essays in Honor of Professor Isamu Yamada
Editors: R. Sato, M. J. Beckmann, Brown University, Providence, RI, USA
1983. VIII, 195 pages.
ISBN 3-540-11998-1

This volume was prepared in honor of the 73rd birthday of Professor Isamu Yamada, the grand old man of Japanese economic theory. 21 Japanese and international contributors submitted papers which can be devided into three categories: microorganization and macroorganization, economic structure, and technology.